T0177688

GRAVITY!

To G.A., G.G., and S.H.

GRAVITY!

The Quest for Gravitational Waves

Pierre Binétruy

OXFORD
UNIVERSITY PRESS

OXFORD
UNIVERSITY PRESS

Great Clarendon Street, Oxford, OX2 6DP,
United Kingdom

Oxford University Press is a department of the University of Oxford.
It furthers the University's objective of excellence in research, scholarship,
and education by publishing worldwide. Oxford is a registered trade mark of
Oxford University Press in the UK and in certain other countries

An earlier version of this book has appeared as *À la poursuite des ondes gravitationnelles*
with Dunod (Paris, 2015)

First Edition published in 2018

Impression: 1

Published in the United States of America by Oxford University Press
198 Madison Avenue, New York, NY 10016, United States of America

British Library Cataloguing in Publication Data

Data available

Library of Congress Control Number: 2017953064

ISBN 978–0–19–879651–0

Printed and bound by
CPI Group (UK) Ltd, Croydon, CR0 4YY

Illustrations by Pierre Binétruy and David Gardner.
Chapter opener image: Simulation of the merger of two black holes and the resulting
emission of gravitational waves. Credit: NASA/C. Henze.

Note

The history of the gravitational Universe is being written rapidly since the recent detection of gravitational waves just 100 years after Einstein's prediction. This major event makes it all the more necessary to have a global vision of the gravitational Universe, as presented in this book.

This compelling discovery pressed Pierre Binétruy to update his earlier French publication, and to include a chapter dedicated to the LISA mission, an experiment that will allow gravitational wave detection in space, and in which Pierre was heavily involved.

Sadly, Pierre Binétruy passed away on 1 April 2017. His early death did not allow him to supervise the final form of this book. With the support of Pierre's family, researchers from the APC Laboratory, his former PhD students, and PCCP fellows and collaborators have united to make the work ready for publication. Thanks to Marie Verleure, Danièle Steer, David Langlois, Valerie Domcke, Dhiraj Hazra, Alexis Helou, Frederic Lamy, Joël Mabillard, and Mauro Pieroni.

Focus VII, Focus VIII, Chapter 11, and the Glossary have been translated directly from the French version. Everything else is Pierre's last writing, including the book layout, the choice of illustrations, and the selected quotes.

Pierre was a great scientific communicator who knew how to transmit the most complex notions of his discipline to the public. The publication of this book was thus very important to him. Considering all the energy and attention to detail he put into it, it has been an honour to help him accomplish at least one of his numerous unfinished projects, on a topic he was passionate about.

Preface

This book is an introduction to the gravitational Universe. It does not require any particular scientific knowledge and avoids all the mathematical formulation that constitutes the backbone of Einstein's theory on general relativity. But it seeks to present its ideas and concepts, those that are well established as well as those objects of the most current research and debates within the scientific community. Understanding them requires personal investment, but all the necessary tools are provided in this book.

The Prologue gives an overview, a road map of what awaits you. Reading can be continuous in a linear manner, from chapter to chapter. You will also find boxes that clarify a notion or more technical information. They can satisfy your curiosity or be left out.

But perhaps you will be impatient to know more. Focuses at the end of each chapter can be read through as you please. They gain, of course, to be read following the associated chapter. However, they concentrate on particular aspects and may be discovered separately.

I have paid particular attention to two important reading tools located at the end of the volume: the Glossary and the Index. The latter links the different themes covered by the book: you will see that the concepts reappear under various lightings that make it possible to better define the contours. In the course of your future readings (I do not doubt that the next few years will be fertile in discoveries in this field), you will quickly find, thanks to the Index, adequate explanations or the scientists who contributed to the development of the ideas and to the discoveries.

This book is therefore also a tool for understanding today and tomorrow the keys of our Universe.

I would like to conclude by thanking all those who allowed me to gradually build this synthesis of our knowledge of the physics of the Universe through their questions and observations: spectators of the various public lectures I was able to give, the 100,000 learners of the free online course 'Gravity!' (French and English versions) with which I had so much pleasure in sharing these exciting hours. Let me also thank Jean-Luc Robert for convincing me to write this book, Marie Verleure for her unconditional support, and Gorka Alda and Gérard Auger for their careful reading. They have been of great help to me.

Pierre Binétruy,

Paris,

December 2016

Contents

Prologue: Transfigured Night

What were you doing on 14 September 2015 at 9 hours 15 minutes 45 seconds Universal Time? That is, 7.15 p.m. in Canberra, 11.15 a.m. in Paris, 10.15 a.m. in London, and 2.15 a.m. in San Francisco.

In order to answer this question myself, I had to check my calendar. I was at the Albert Einstein Institute in Hanover attending a symposium on gravitational waves. This conference was organized by our German colleagues to discuss collaboration with China on a future space mission designed to study gravitational waves, those elusive waves often described as 'ripples of space–time across the Universe'.

Just at that moment, one of these powerful gravitational waves was deforming Earth and all objects on it. It was a nonevent on a cosmic scale, because gravitational waves are ubiquitous in the Universe. They regularly run through even our small Earth. This was a nonevent for you and me because we absolutely do not feel them: their effects on material bodies are incredibly tiny. But it was a truly historical event for mankind, because, for the first time ever, a detector built by man, LIGO, identified the passage of one of these waves.

Little did I know at the time that, in the very building where I was, on that same day, surprise then excitement were reigning. The Albert Einstein Institute is part of the LIGO collaboration. The detector had just started to take in data in its advanced version when the automated data analysis identified a strong signal. 'Too good to be true', but the first human checks showed that this signal had the exact form expected of a gravitational wave produced when two black holes merge. The only surprise was the strength of the signal. But, at the end of the day, most members of the collaboration knew this was, to all probability, the gravitational wave discovery they had long been waiting for: one hundred years after Einstein's article had predicted their existence!

The stakes were so high that announcing such a discovery required exceptional care. Moreover, the signal was so clean that the physicists had to extract all the science from it. And what a science! The science of black holes had fascinated so many over the years: scientists, philosophers, science fiction writers, ... basically all of us. Many months were spent checking all the details, months when rumours of a major discovery appeared, disappeared, reappeared.

I don't have to check my calendar to know what I was doing five months later on February 11. I was in a small hotel room in Geneva, Switzerland, eagerly watching the press conference held at the National Science Foundation in Washington, DC. The suspense was intense but brief: 'We ... have discovered ... gravitational waves.' It was a very emotional moment. I was almost glad that I was all alone in my room.

That night at 10.30 p.m. I was interviewed live by a French information TV channel. A new French cabinet had been announced that same day, and I had been warned that this would make most of the 10-minute news slot, but they would start nevertheless with the discovery. I knew my message had to be short, so I thought about what I wanted to get across: this discovery is historic because it concludes a 100-year-old search, but even more so because it opens a new era in the exploration of the Universe, of the mysterious gravitational Universe.

Well, I was able to say much more than that, because the interview lasted 6 minutes. And at the end of these 6 minutes, the journalist said: 'Well, I know that, after all that we have just heard, this will sound so irrelevant. But we also have to cover the announcement of the new French cabinet.' I suddenly got the impression that, for a very brief moment, this tiny 'ripple' had brought everything to the right perspective. Oh, a very brief moment! But it felt so good.

> Tâchez de garder toujours un morceau de ciel au-dessus de votre vie.
>
> *Always try to keep a piece of sky over your life.*
>
> MARCEL PROUST, *Du côté de chez Swann* (1913)

Let me now take you outside, on a beautiful moonless night. We have all marvelled at the scintillating display of stars speckling the sky. After observing for a few minutes, we can locate the opal band of the Milky Way and some constellations; we can identify various shades of light and make out a swarming in the background, guessing the presence of even more distant stars. The Universe lies before us in all its glory and immutability. And we feel like specks of dust thrown among these grains of light, for an instant pathetically small compared to the cosmic ages. For centuries, poets, philosophers, artists, and storytellers have dreamed, written, and debated about this dizzying perspective offered to anyone who raises their head to contemplate the starry night sky. All has been said about it. Yet …

Yet the physics of the past century has taught us that the observable Universe is much richer than what may be revealed to our eyes. This richness is accessible now, accessible through means of detection more powerful than our eyes, as we have known since the days of Galileo and his telescope. It is also accessible through tools which do not probe light but other types of radiation. And we are only starting to understand the consequences of this, and are reaching for the more fundamental nature of the Universe.

Let us return to our starry night. Each grain of light is colloquially called a star, yet we have known for a long time that some of them are the planets of our own Solar System, reflecting the light of the Sun: Mercury, Venus, Mars, Jupiter, Saturn, Uranus, and Neptune. We have also known for a century that a few of these luminous dots are galaxies, enormous accumulations of stars. Our own star, the Sun, is part of such an ensemble, the Milky Way, our own Galaxy. Because the Solar System is somewhat at the outskirts of this Galaxy, we see the Milky Way from the side: hence, its elongated shape across the sky. The milky aspect of this strip is due to the concentration of stars in this direction. We also see Milky Way stars in other directions, but they are more scattered. All the stars that we can identify in the sky are actually part of our Galaxy. A few luminous dots are of *extragalactic* nature, which means that they are localized outside our Galaxy. They are, in fact, galaxies in their own right: they are located so far away that we cannot separate the individual stars that form them.

The discovery of the extragalactic nature of these light sources, i.e. the discovery of other galaxies, in the 1920s came with an even more surprising finding: these galaxies move away from us. Thus, the Universe, which had impressed us with its immutability, is on the contrary dynamical. And since all galaxies recede from us, we must deduce, if we are not in a privileged location, that it is the very structure of space–time that is dynamical: each galaxy moves away from every other galaxy. We translate this by saying that the Universe is expanding.

Light provides the proof of this expansion. Indeed, we may study the material elements present in each star by analysing the light they emit, in particular the colour of the light associated with its frequency. The work of Lemaître and Hubble in the 1920s showed that the light emitted by known elements in the cosmos is slightly shifted in frequency towards the red, an effect that would later be attributed to their motion away from us. The distant galaxies that emit them are receding from us.

But light also has a finite velocity. And this has truly remarkable consequences for us observers of a starry night. The light emitted by a celestial body takes a finite amount of time to reach us: 8.32 minutes from the Sun, 100,000 years from the borders of the Milky Way, 2.5 million years from the Andromeda Galaxy. This means that, when we contemplate the sky, we are not watching the Universe as it is today, but rather a series of pictures that are all the more ancient that they are more distant. A sort of static movie unfolds in front of our eyes.

It is a formidable advantage for anyone interested in the Universe's history: this history is narrated for us, here and now. The most distant celestial bodies that we see (located some 14 billion light years away) are in the state that they were in at the beginning of the Universe. What has happened to them since? What is their present state? In order to find out, we need to wait another 14 billion years for the light that they emit now to reach us. Conversely, an observer from the Andromeda Galaxy would see today Earth as it was some 2.5 million years ago, more or less at the time when *Homo erectus* appeared.

Thus, the light of distant stars and galaxies provides us with a slice across space and time. From it we reconstruct the Universe in its spatial and temporal entirety, just as the rings in the cross section of a freshly cut trunk allow us to reconstruct the history of the tree and of its environment.

But what is responsible for the Universe's expansion? Where does the energy necessary to dilate these distances come from? The answer has been provided by Einstein. In 1915, one hundred years ago, he was looking for a system of equations that unified the description of gravitational phenomena in the framework of his theory of special relativity conceived ten years earlier. These would become Einstein's famous equations for quantifying the deformation of space–time under the effect of a mass, or more generally of a concentration of energy. They make it possible to compute the trajectory of an object close to a massive body. And they form the heart of what is called general relativity.

Soon Einstein attempted to apply his equations not just to objects (such as celestial objects) in gravitational interaction, but to the whole

Universe. And indeed, solutions in which the Universe is in expansion exist. The origin of expansion must be searched for in the laws of gravity itself, which tells us that any form of energy induces the expansion. In a sense, the Universe as a whole is run by gravitation! Paraphrasing Aristotle, according to whom God was the first engine of the Universe, we see that *gravitation is the first engine of the Universe's evolution*. This leads us to question the status of gravitational force among fundamental forces. It is the first force that we become aware of: if possible, just recall your first efforts as a toddler to stand up and walk. But it is also the weakest of all fundamental forces. What is the reason for this apparent paradox? The force of attraction is proportional to the mass and we live close to a very massive object, Earth. Chances are high that, had we lived on an asteroid, we would have paid less attention to gravity! Now, because it is weak, it is the least known of the fundamental forces. We know that it has a very long (in principle infinite) range: this is why it runs the Universe at large. But, at short distance, it is much less known than the other long-range force, the electromagnetic force.

Actually, over the course of the twentieth century, the theory of gravitation has developed rather independently of other interactions. Whereas electromagnetism has made it possible to study the microscopic Universe, for example by projecting charged particles onto matter, and to unravel the laws of quantum mechanics, general relativity has successfully passed a certain number of striking confirmations in the nearby Universe, namely our Solar System (e.g. curvature of light rays close to the Sun, movement of planets).

Exploration of the infinitely small continued throughout the twentieth century with the discovery of the nuclear or strong force and of the weak force responsible for certain types of radioactivity, and the development of increasingly powerful accelerators to study them. Since particles travel in accelerators at velocities close to the speed of light, quantum mechanics had to be reconciled with special relativity, which was achieved in the years 1940–1950.

It appeared later that the concepts developed for electromagnetism could be generalized to describe the weak and strong forces. This led to the unified theoretical description known as the Standard Model. The final confirmation of this Model was the discovery of the Higgs particle at CERN in 2012, the last elementary brick necessary for the consistency of the Model.

General relativity, the theory of gravitation, has developed independently in parallel, mainly through theoretical works on singularities.
These singularities are 'pathological' behaviours of the theory either in
a distant past (the famous Big Bang) or in the gravitational collapse of a
star (the equally famous black hole). Even if some of its phenomena,
such as Hawking radiation emitted by black holes, have some quantum
aspects, gravity still escapes a full quantum treatment. Is the theory of
gravity inherently irreconcilable with quantum theory, or should we
modify the laws of quantum mechanics to make room for gravitation?

In the years 1970–1980, a bridge was established between the two theories. On the side of the three nongravitational forces, it was the march
towards grand unification and the possibility that the weak, strong, and
electromagnetic forces are, at low energy, different facets of a unique
force that could only be accessed in its entirety at very high energy. On
the gravitational side, we became gradually aware that the further back
in time we went, the hotter and denser the Universe became: ultimately,
it boiled down to a hot soup of elementary particles. But the intensity
of the gravitational force increases with energy; consequently, in the
first instants of the Universe, gravitational interaction between elementary particles was as strong as the other forces and should be governed
by the laws of quantum mechanics. Will all known fundamental forces,
including gravitation, eventually be determined to be different aspects
of a single elemental force?

An enduring problem, however, highlights the difficulty of reconcil-
ing gravitation with quantum physics. It is the problem of *vacuum energy*.
In quantum theory, each system has a ground state, or state of lowest
energy, from which higher energy states are built. It is colloquially
called the vacuum. The Universe itself has its own vacuum state. This
quantum state is a medium besieged with (quantum) fluctuations that,
on average, contribute a nonzero value to the energy. But, in the context of general relativity, each form of energy deforms space and time,
and is thus in principle measurable. This must be true for the vacuum
energy as well. A back-of-an-envelope computation using characteristic quantities of quantum physics and general relativity predicts a
ridiculously large value (120 orders of magnitude too large) compared
to what is actually compatible with observation. Thus, *a major difficulty
arises when we attempt to unify gravitation with the other fundamental interactions.* The
ultimate unification will certainly require the two theories, general

relativity and quantum mechanics, to be reconsidered in a fundamental manner. At the end of the 1990s, the general feeling was that, once the unifying theory was found, it would lead to the automatic cancellation of vacuum energy.

However, a nonzero vacuum energy is necessary to explain the rapid (exponential) expansion phase the Universe experienced right after the Big Bang. It is this phase, called *inflation*, that enables us to explain why the Universe we observe today has identical properties in all directions: it originates from the expansion of a tiny region of space–time.

The Planck mission of the European Space Agency, launched in 2009, was the last in a series of increasingly precise observations to support the predictions of inflation models. It appears that the structures observed in the cosmological background, the first light emitted in the Universe 380,000 years after the Big Bang, are the relics of fluctuations emitted within the quantum vacuum during the inflation phase.

Meanwhile, vacuum energy moved centre stage thanks to the discovery in 1999 of an *acceleration of the Universe's expansion* in its recent history. Explosions of a type of supernova, known for the stability of its brightness, seemed to be less bright when the supernovae are older. This was interpreted as proving that the supernovae are located further away than anticipated, which means that the Universe has expanded faster than expected since the supernovae explosions: it has undergone acceleration. The most remarkable aspect of this discovery lies in the fact that almost all known forms of energy of the Universe (matter, radiation) tend to decelerate the expansion. A new type of energy must be responsible for this acceleration. New? Unless it is vacuum energy, which is the only known form that induces an acceleration.

Is vacuum energy the main energy component of the Universe driving the recent acceleration of its expansion? A very large observational programme has been set up on the ground and in space in order to try to answer this question. It would be ironic if we eventually get the confirmation that this form of energy is dominant. This is another reason why the reconciliation of gravity and the quantum theory is at the top of the fundamental theory community's agenda.

However, in the long series of discoveries that we have reviewed, the most extraordinary was yet to come!

Just as an electric charge in motion generates an electromagnetic wave, the rapid displacement of mass (as in an explosion, for example) produces a curvature of space–time that propagates, just as the fall of an object into a pond generates ripples on the pond's surface. Those are the gravitational waves predicted by Einstein just one year after he published his work on general relativity. When these waves pass through the laboratory, the corresponding curvature of space–time appears to set in motion distant objects with respect to one another.

Because the gravitational force is very weak, this effect is extraordinarily small, so small it was believed to be practically impossible by Einstein and the first researchers. The first who took up the challenge, J. A. Wheeler and J. Weber, were considered daydreamers. And it took many years of dedication, insight, and technological and conceptual achievements to meet a goal that had seemed for so long unreachable: the detection of a gravitational wave. And we are back to February 11, the day the discovery's announcement was made.

It is amazing to realize that on that day, the world was already geared towards further exploration of the gravitational Universe. Several ground detectors had been built around the world; there was a detailed plan for a space mission. This means that a large community of some 2,000 scientists was so firmly convinced about the importance of gravitational waves in understanding the Universe that they had decided to devote their scientific life to this task, without having detected a single one!

Whereas for centuries it has been light that has allowed mankind to observe and understand the Universe, we are now ready to observe it through waves of gravitation, this very gravitation which, as we have learned, is the first engine of the Universe.

It is to a better understanding of this key moment in the evolution of the most fundamental concepts of our Universe that I invite you. It is a key moment because it is the result of 100 years of thinking about the theory of general relativity, the first global theory of gravitation. But also for the reason that a number of major discoveries, all connected with gravitation's role in the Universe, have taken place during the past fifteen years. Finally, it is a key moment because a global observation programme which is expected to revolutionize our understanding of the gravitational Universe has been set up.

The most fundamental questions are tackled by these theories and observations. It is important that each of us is aware of these questions, because they are likely to change our perception of our place in the Universe. This is why this book is intended for those of you who are curious about these issues, whether you have a scientific background or not. You may come across notions or concepts unknown to you, and the unknown often appears to be complex, but one of the roles of the physicist is to introduce order into this complexity. After all, cosmos means order in ancient Greek!

1

Gravity's Rainbow:
Galileo, Newton, Einstein

Watching objects fall gives us a direct experience of gravity, of the gravitational force that attracts any massive object to Earth. This is the force the toddler must reckon with in order to stand on its feet. It is probably this urge to stand up and fight against gravity that partly defines our uniqueness as human beings. It is also from experimenting with gravity that the concept of force has emerged in modern sciences.

And yet the force of gravity is the weakest of the four known fundamental forces. It is directly perceptible only for a fortuitous reason: the presence of a massive body in our immediate neighbourhood, the Earth. Imagine for a second that we were born in weightlessness, far from any star like the Sun or any planet like the Earth. The first force that we might experience would probably be a frictional force of some kind, such as the resistance one encounters by turning a spoon in a jar of honey. Of which nature is this force? It is electrical: friction results from the bonds (electrical in nature) that exist between the atoms of the moving body (the spoon) and those of the surrounding environment (the honey).

Let me clarify the concept of *fundamental* force. A fundamental force is described by a law that has the same form at any point in space and at any time, meaning that it is valid throughout the Universe and throughout its history. Understanding such forces is thus key to understanding the principles that govern our Universe, not only today but also in the past and in the future. Fundamental forces are also called *elementary forces* because, strictly speaking, they are exerted between the *elementary* constituents of matter: the electric force or Coulomb force acts between two elementary charges, e.g. between two electrons. Because of the reciprocity of the action between two elementary constituents, physicists often talk of *elementary interactions*.

All forces between two bodies are not elementary: they are often the combination of multiple elementary forces acting between their elementary constituents. Therefore, they do not obey fundamental laws, but empirical approximate laws, and they are not necessarily applicable everywhere and always. A good illustration is frictional forces. Imagine that we are pushing a crate over a rough surface: the force of friction that opposes the motion is resulting from the multiple elementary electric forces between the electrons of the crate and those of the surface (these forces generate bonds between the molecules of the box and those of the surface). This force will depend on many conditions: roughness of the surface, roughness and shape of the crate, velocity of the crate, temperature of the room, etc. The law describing the force resulting from these multiple parameters can only be empirical.

In contrast, the electric force, or Coulomb force, is a fundamental force; as a matter of fact, since the late nineteenth century when the description of electric and magnetic phenomena was unified, we speak of the electromagnetic force. Nowadays, we identify four fundamental forces: besides the two just mentioned (electromagnetic and gravitational), we count the nuclear force or strong force, and the force associated with certain types of radioactivity known as the weak force.[1]

A brief history of how physicists have come to consider gravitational force as a fundamental force provides insight into the profound nature of this force. It will also allow me to introduce various concepts that will accompany us all throughout this book.

If I drop a ball of lead and a feather, they will not fall in the same way. If I repeat this experiment in water or oil, the movement of the ball will be different … and the feather will float. So there is an influence of the medium, air, water, oil, on the falling motion. Aristotle deduces that the medium is necessary for the movement and thus rejects the idea of void: the movement would be instantaneous in vacuum, which is absurd. Actually Aristotle distinguishes two types of movements: the natural motion (up or down, depending on whether the body partakes more of the air or of the earth), and forced motion (under the action of a force). The movement of stars, which seem neither to fall nor to move away since they reappear every night, is in a special category: the circular motion, perfect and eternal, of celestial spheres. And, according to

---

[1] We return to these two forces in Chapter 4.

Aristotle, God is the primal engine, necessarily motionless, that causes the movements throughout the entire Universe, the 'unmoved mover'.

The birth of modern physics is usually traced back to Galileo, especially for his use of experimentation to test each hypothesis, actually both real and virtual experimentation. I will give you some examples of his thought experiments. This will allow me later, when we reach more elaborate concepts not easily tested in real life, to probe them with carefully designed thought experiments that will allow me to outline more precisely their meaning and their limitations.

Galileo and the universality of free fall

Our story starts in 1583 in the cathedral of Pisa. According to the legend, 19-year-old medical student Galileo Galilei is observing the oscillations of the central chandelier: measuring the period of oscillations with his own pulse, he realizes that this period does not depend on the amplitude of the oscillations. Back home he checks this observation with pendulums of his own making and notices as well that the period does not depend on the mass of the hanging object! There is some universal character in the motion of a pendulum. This leads him to consider that the motion of objects in the gravity field of the Earth may be more universal, and thus more fundamental, than previously thought by Aristotle.

To investigate further, Galileo realizes that one of the difficulties of an experimental study of falling objects is the brevity: a fall from a height of 5 meters lasts about one second.[2] He thus decides to use a simple setup in order to slow it down: he lets objects slide along an inclined board: the smaller the angle of the board is with the horizontal, the slower the motion is. You could argue that he is increasing friction, but by testing different types of surfaces (rough, smooth, waxed, iced), he infers the effect of friction and imagines an ideal board with no friction. Galileo concludes from his experiments that the motion is universal (the same for all bodies) and is uniformly accelerated, which means that the distance covered varies with the square of time. This is a characteristic of the motion of a body on which a constant force is exerted.

---

[2] The fall from the famous Tower of Pisa lasts less than 4 seconds.

Uniformly accelerated motion

In order to check that the motion of an object falling along an inclined board is accelerated, one may trace five equidistant marks on the board identified by the numbers 0 to 4 from top to bottom. Release a ball at position 0 without initial velocity and measure the time it passes position 1 and then position 4. If the ball took twice as much time to go from 0 to 4 as from 0 to 1, i.e. covering a distance 4 times larger, the motion is uniformly accelerated: the distance travelled (4) varies as the square of time (2).

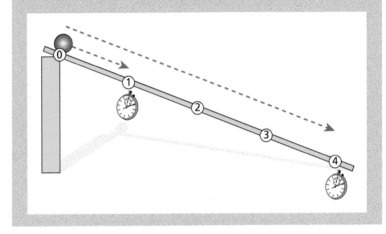

In order to better judge the effect of friction I can slightly complicate the setup by provoking an upturn of the object motion with a second inclined board B (Figure 1.1a). The less friction there is, the higher the object rolls upward. In the ideal case where there would be no friction, the object would reach the same height as its initial position. In this limit, the acceleration (obtained by dividing distances travelled by the square of the corresponding time) of the descending motion along board A tends to a constant, about 10 m/s², the acceleration of Earth attraction.

The less inclined board B is on the right-hand side, the further the object reaches to recover its initial height (Figure 1.1b). If I make the board horizontal (Figure 1.1c), the object should continue its motion at constant velocity because it can never recover its initial height. In the ideal limit when friction is vanishing, this motion should continue indefinitely at constant velocity.

This allows us to define a characteristic of any material object: its inertia, that is its capacity to resist any variation of motion. According

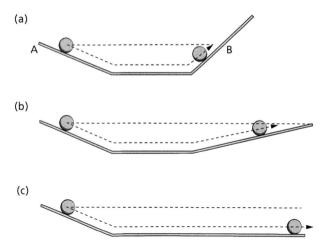

Figure 1.1 From the inclined board experiments of Galileo to an illustration of the principle of inertia. In the absence of friction (thought experiment) the ball returns to the same height. The more inclined board B is, the further to the right the ball travels (a, b). If the board is horizontal (c), the ball continues indefinitely at constant velocity.

to the *principle of inertia*, an object with no net force acting on it has a uniform motion, i.e. a motion at constant (or vanishing) velocity.

This principle of inertia is a consequence of another principle stated by Galileo, the principle of relativity. Aristotelians considered that Earth must be at standstill; otherwise, they said, a ball thrown vertically upward would not fall back to the same place because meanwhile the Earth would have moved. Galileo noted that if you are isolated in the cabin of a ship, with no possible view outside, you cannot make an experiment that would tell you whether the ship is at standstill or in uniform motion: the motions that you observe, for example that of a ball thrown in the cabin, are identical in the two cases. The Galilean principle of relativity thus states that the results of experiments, and hence the laws of physics, are identical in all reference frames (such as a ship) in uniform motion with respect to one another.

One can easily imagine that the inertia of a body depends on the quantity of matter that constitutes it. It is, in fact, the mass of a body that characterizes this inertia. A practical example will allow you to understand why. Imagine that you must set in motion a mine cart. You would certainly prefer to do it with an empty cart than with a cart full of ore. Why? Naively, we might think it is because of its weight that the loaded cart presses more forcefully on the rails, but the motion is horizontal and the weight is entirely compensated by the reaction of the rail. It is, in fact, the quantity of matter, measured by the mass, that determines the resistance to motion, in other words the inertia. In order to check this, try now to stop the cart in motion. If you thought that the weight on the rail was acting against you before, this time it should help you. But it is the empty cart that is easier to stop because its mass, and thus its inertia, is smaller.

The difficulty resides here in the confusion made in everyday language between mass and weight. But the physicist must, on the other hand, distinguish between force, i.e. the weight, measured in Newton, and the inertia, i.e. the mass measured in kilogram. A last example: an astronaut in weightlessness on a space station wants to know whether a packet of sugar is empty or full. In order to do that, she moves it around: the more the box resists motion, the more sugar it must contain.

Let us return for a second to the chandelier of Pisa Cathedral. Galileo rightfully guessed that the same law governs both the oscillation of a pendulum and the fall of an object in Earth's gravity field. You may note in Figure 1.2 the analogy between the motion of a ball along a board, which we considered earlier, and that of a pendulum. The pendulum

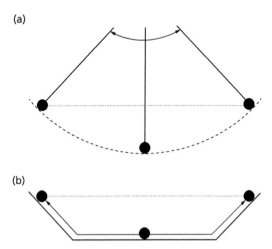

Figure 1.2 Comparing (a) the motion of a pendulum and (b) the motion of a ball between two inclined boards.

rope counteracts the fall and it makes it possible to significantly decrease the friction: the only friction comes from the ambient air.

Newton and the Moon falling

If we owe Galileo the definition of the concept of force from his detailed study of falling material bodies in Earth's gravitational field, it is Isaac Newton who identified the gravitational force as a universal force that governs the motion of stars and planets. He was preceded by Johannes Kepler, who had identified the laws of the planetary motion, however without explaining the causes.

According to what Newton said, it was the unlucky fall of an apple on his head that started his thoughts on the subject. Let us return to our fictitious experiments on falling material objects to understand what is common between the fall of an apple and the motion of stars. Let us climb a ladder and drop an apple from a height of 5 m without initial velocity (Figure 1.3a). We see that it reaches the ground one second later.

If we give this apple a horizontal velocity of 1 m/s, it will fall 1 meter from us: one can indeed decompose the motion into a vertical motion (identical to the previous motion and taking one second) and a horizontal motion (the apple moves a horizontal distance of 1 meter during this second). If we give the apple a velocity of 2 m/s, it will fall at a distance of 2 meters. With a velocity of 7900 m/s, it should fall at a distance of 7900 meters from us. Well, in fact, not exactly! This would be true if the Earth was flat but it is round and thus its surface is curved: I have chosen the distance in such a way that by falling a distance of 5 meters along a trajectory 7900 meters long, the apple still finds itself at a height of 5 meters from the ground, if we can neglect the air friction. In other words, the apple has followed the curvature of the Earth and, without friction, would continue indefinitely (by inertia): we have put the apple into orbit! This is the *thought experiment* called Newton's cannon.

Of course, the experiment cannot be realized in practice because friction cannot be neglected: at this velocity, friction on the air would completely annihilate the apple. But this thought experiment shows us that the falling motion of the apple with initial velocity is not different from the motion of a satellite in orbit (Figure 1.3b) or even of our natural satellite the Moon. The Moon is thus in constant free fall but with a horizontal velocity such that this fall translates into a circular motion

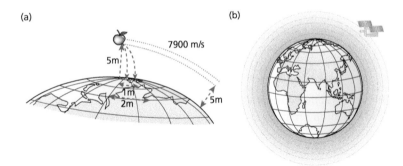

Figure 1.3 Experiment of *Newton's cannon* (not to scale!). (a) An apple thrown at an initial horizontal velocity of 7900 m/s would still be at the same height (5 m) from the ground one second later, or at any later time. (b) The apple is thus put into orbit, like the International Space Station.

around the Earth. This generalizes to planets in free fall around the Sun and to all stars. We conclude that the motion of stars is in essence not different from the motion of an apple, despite what Aristotelians believed.

It is also Newton who understood the characteristics of the force responsible for the universal attraction of material and celestial bodies, the gravitational force. This force between two bodies depends on two pieces of data: the quantity of matter (measured by the mass) that is present in either body, and the distance between them. The larger the distance is, the weaker the force. More precisely, according to *Newton's law*, the intensity of the force is proportional to the mass of each body and inversely proportional to the square of the distance. The constant of proportionality, called Newton's constant, is universal because the force is fundamental: it has the same value everywhere and at every moment. It was only measured precisely in 1798 by Henry Cavendish and its value is 6.66×10^{-11} Nm^2/kg^2. This means, for example, that the gravitational force exerted between two masses of 1 ton (1000 kg) placed at a 10-meter distance is 6.6×10^{-7} N. The Newton, noted N, is the unit of force. Unfortunately, it is not used in everyday life and this may not tell you much. But, believe me, a force of 6.6×10^{-7} N is extremely small: the weight of a mosquito that lands on one of these masses represents a force 10 times more important! This shows how small the gravitational attraction between two masses is.

According to the laws of motion, again established by Newton, if a body exerts a force on another body, then the latter exerts on the former a force that is equal in strength and opposite: this is the *principle of action and reaction*. Thus, the gravitational force is exerted by the Sun on the Earth as well as by the Earth on the Sun. One is speaking of gravitational *interaction*.

Let us come back for a second to our example of the apple falling. Since the origin of the motion is identical to the one of planetary motion, the force exerted is the gravitational force, that is the interaction between the Earth and the apple. One deduces that the Earth is falling on the apple in just the same way as the apple is falling on the Earth, but it is the apple that is set in motion because it has less inertia (mass).

Einstein and space–time

Let us jump a few centuries to near the beginning of the twentieth century, in 1915, the year in which Albert Einstein revolutionized the study of the gravitational force with his theory of general relativity. But let us concentrate first on the preliminary step that was special relativity, conceived by Einstein ten years earlier, in 1905, and to which are associated as well the names of Lorentz and Poincaré.

For this we must return to Galileo and his principle of relativity, according to which the results of an experiment, and thus the laws of physics, are identical in all reference frames *in uniform motion with respect to one another*. They are called, in his honour, Galilean reference frames. Galileo used the example of a ship moving at constant velocity; as a tribute to the twentieth century that had just begun, I will take in what follows the examples of a train, of a lift, and of a rocket.

What is a frame of reference?

Since time and space play a key role in what we will be discussing, it is crucial to define precisely the place and time where one will make an experimental measurement. In order to do this, one must define a reference frame which will make it possible to map space and time. One can imagine this reference frame as a grid in the three dimensions of space: at each grid site, a clock measures the time of events at this point (see Figure 2.1b).

One of the scientific revolutions that took place between the times of Galileo and Einstein is the discovery and characterization of electromagnetic phenomena, culminating in James Clerk Maxwell's unified description at the end of the nineteenth century. Maxwell summarized all known electromagnetic phenomena with a set of equations where the fundamental parameter was the velocity of light, which is not surprising because light had been identified as an electromagnetic wave. The question then raised is whether one should extend the principle of relativity to electromagnetism: if the results of experiments are the same on a train in uniform motion and in a train station at rest, then the physicist on the train and the one who remained in the station must measure the same velocity of light, a fundamental constant. This seems to collide with our natural understanding of the addition of velocities; yet the experiments of Albert Michelson and Edward Morley between 1881 and 1887 had concluded accordingly that the velocity of light measured on Earth is the same in all directions, whereas Earth is in motion with respect to distant stars. In other words, *the velocity of light is the same in all Galilean reference frames: it is a fundamental constant.*

The fact that the speed of light should not be added to the velocity of the reference frame where the measurement is made has some unexpected consequences on the very notions of space and time. These consequences are at the basis of the theory of relativity and of its most surprising properties, at least for our common sense. Let me illustrate this with a simple experiment on board a train. A physicist, say Einstein, has boarded the train to measure the velocity of light during the ride. He does so by sending a light ray vertically between the floor and ceiling of the carriage: by measuring the ratio of distance covered over time he obtains the velocity of light. The cow Margie peacefully watches from a clover field the train pass her at constant velocity and, through the carriage window, the experiment being performed. She carefully notes the times of departure and arrival of the light ray (Figure 1.4). Since the train is moving with respect to her, the light detector has been moving during the time interval and the distance covered by the light ray is seen by Margie to be larger than the one measured by Einstein on the train. Since both Einstein and Margie must measure the same velocity of light, we conclude that the times measured are different. Time runs faster in the train where Einstein conducts the experiment![3]

This unexpected effect was measured experimentally by comparing two sets of clocks, one of which had travelled for many hours in two

---

[3] Time runs slower—it is *dilated*—in the reference frame of Margie's clover field, which is in *relative motion* with respect to the measurement device on the train.

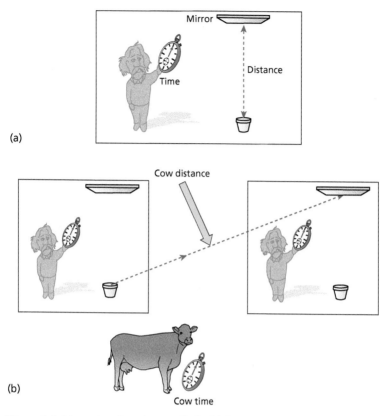

(a)

(b)

Figure 1.4 Measuring the velocity of light (a) from inside a train and (b) from the field that a train wagon is passing at constant velocity.

planes, whereas the other had stayed motionless.[4] It is also verified everyday in accelerators, where the particles move at velocities close to the speed of light: the effect is all the more important that the velocity is close to the speed of light. And, needless to say, it is completely negligible in a train: we were just performing a thought experiment!

But more importantly this challenges our innermost perception of time and space by unifying them into a single concept, space–time. In the experiment that we have just imagined, we singled out the role of the time dimension, but similar effects are observed on spatial dimensions:

---

[4] Experiment conducted by physicist Joseph C. Hafele and astronomer Richard Keating in October 1971 (*Science* 1972).

the *dilatation of times* in moving reference frames corresponds to the *contraction of lengths* in such frames. These surprising effects have made the theory of special relativity popular. An amusing illustration was proposed by the cosmologist George Gamow, who imagined, in the *New World of Mr Tompkins*, everyday life in a village where the velocity of light is 32 km/h: a simple cyclist would be witnessing spectacular relativistic effects associated with the dilatation of times and the contraction of lengths.

$E = mc^2$, more than an equation

The most famous equation in physics is not the most important in the theory of relativity, but it ended up by symbolizing the genius of Einstein. This equation tells us that a mass (m) is a form of energy (E), which is an important realization of *special relativity*. The parameter c is nothing else than the speed of light that, as we have seen, plays a central role in this theory.

Einstein and the fall of the lift

It is time to address the theory of general relativity and to explain in which sense this theory enabled Einstein to conceive a theory of the gravitational force. The generalization referred to here is linked to the fact that the theory of relativity of 1905 was *special* to Galilean reference frames, i.e. a class of frames in uniform motion with respect to one another. This limitation was a preoccupying subject for Einstein: what form do the laws of physics take in a reference frame that is not in uniform motion but in accelerated motion? Are they different from those in a Galilean frame and, if so, how different?

This question drew Einstein into a domain that was a priori unexpected: gravity. In order to understand this, let me come back to the principle of relativity: after the ship of Galileo, the train of Albert Einstein of 1905, let us shut ourselves away in a spacecraft in uniform motion (constant velocity), far from Earth and any star. We are in weightlessness: if we drop a ball, it will float around us. Imagine now that the spacecraft fires its rockets to develop an upward acceleration of 10 m/s^2: its velocity is no longer constant and we are projected onto the cabin floor, just like the ball. Everything happens as if a gravitational force were pulling us to the ground, and the motion of the ball is identical to that observed in Earth's gravity field. But there is no gravity!

Einstein elevates this observation to the level of a principle. He postulates that, if we are not in visual (or any other means) contact with the exterior, no experiment performed inside the spacecraft allows us to identify whether the observed motions inside the cabin are due to the accelerated motion of the spacecraft (our reference frame) or to the presence of a gravitational field of some nearby star or planet. If you think a little, you will realize that, underneath this postulate, there is the fact that the mass that characterizes the resistance to changes of motion (and hence to acceleration) is identical to the mass that characterizes the quantity of matter in Newton's law for gravitational force. This is known as the *equivalence principle* between inertial mass and gravitational mass.

This postulate can be used in reverse. Imagine, for example, that, back on Earth, we are going down a lift. The cable breaks and the lift goes into free fall. During the time of the fall, we are floating in the lift and if we drop a ball it will appear floating as well (because it is falling with us). Einstein's postulate tells us that during this fall we cannot conduct a physics experiment that would tell us whether we are in free fall in Earth's gravitational field or whether we are in weightlessness far from the gravity field of any massive star or planet.

This equivalence between an accelerated reference frame (a rocket, a lift) and a gravitational field allows us to understand simply some of the most well-known consequences of general relativity. I will only give here a single example. Let us get back to our rocket in uniform motion and send a horizontal light ray across the spacecraft cabin. We know that light propagates in straight lines, at least in Galilean reference frames where we have experimented with it (Figure 1.4). If the spacecraft fires its rockets to accelerate, it will be displaced imperceptibly further upward during the time that the light ray progresses across the cabin. It follows that, for an observer inside the accelerated cabin, the light appears to propagate in a line that is slightly curved (Figure 1.5).

You may think that all this is only a question of individual perception. But let us use Einstein's postulate. Since we cannot fundamentally tell the difference between an accelerated frame and a gravitational field, we can deduce that light is following a curved trajectory in a field of gravitational attraction, which means the light is close to an important mass. This effect was experimentally verified by observing, during the solar eclipse of 1919, the slight curvature of light rays that graze the surface of the Sun.

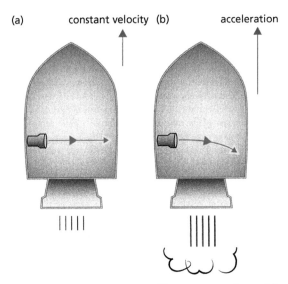

Figure 1.5 Light propagates along curved lines in an accelerated frame, hence in a gravitational field.

This observation is really what triggered the legend of Einstein: this young theorist had an incomprehensible, apparently implausible theory predicting that light beams are bent by mass, contradicting what had always been taught in school, yet nature seemed to confirm it (Figure 1.6).

Let us retain from all this that massive objects curve the geometry of space–time. If you consider, for example, the room where you are reading this book, you may map this room with light rays by setting at regular intervals laser beams along two adjacent walls of the room. These light rays follow the shortest path between two points and thus propagate in straight lines; they also cross one another at right angles (Figure 1.7a). Let us now imagine that your neighbour is concealing a very massive object just behind one of the walls (it must be very massive: say, a black hole a few meters in size!). The trajectory of light rays in this part of the room closest to the partition wall is deformed: they no longer cross at right angles. They still follow the shortest path between two points but no longer in a straight line (Figure 1.7b). Space–time is curved in the vicinity of this large mass.

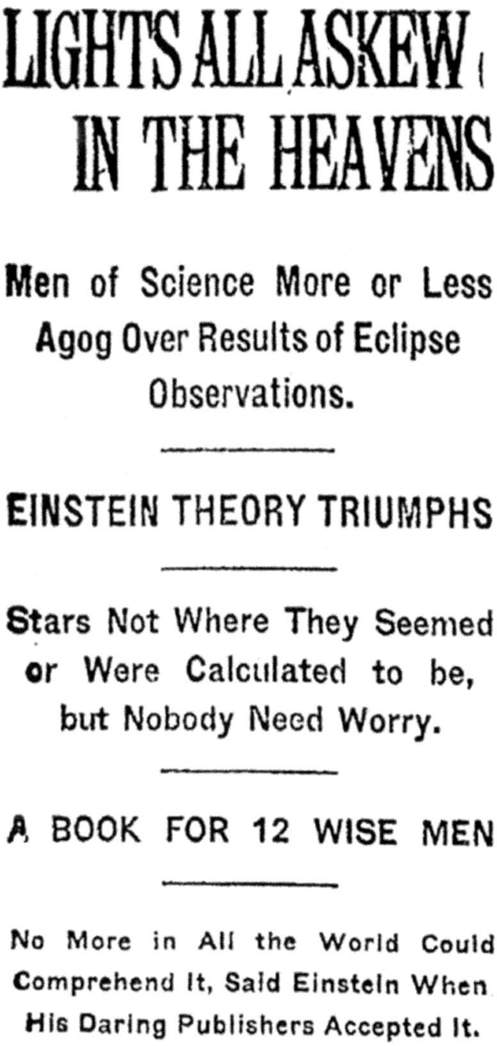

LIGHTS ALL ASKEW
IN THE HEAVENS

Men of Science More or Less Agog Over Results of Eclipse Observations.

EINSTEIN THEORY TRIUMPHS

Stars Not Where They Seemed or Were Calculated to be, but Nobody Need Worry.

A BOOK FOR 12 WISE MEN

No More in All the World Could Comprehend It, Said Einstein When His Daring Publishers Accepted It.

Figure 1.6 Headline of the *New York Times* on 10 November 1919.

Einstein's equations summarize this link between the geometry of space–time and the distribution of mass, or more exactly any form of energy. Indeed, we recall the famous equation $E = mc^2$ that tells us that any mass is a form of energy. Conversely, any form of energy will also have effects on the geometry of space–time.

Thus, a careful analysis of the principle of relativity, initiated by Galileo, led Einstein to develop a theory of gravitation and more generally a theory of space and time that would revolutionize our vision of the Universe.

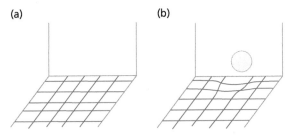

Figure 1.7 A black hole behind a partition wall. A very massive object, such as a black hole, placed behind the wall of a room would slightly deform the straight trajectory of light rays used as a grid to map the room. This would make it possible to identify the presence of this massive object without visiting the neighbour.

Experimenting with gravity

We have all been experimenting with gravity, probably from the day of our birth when we left the maternal placenta.

The toddler who attempts to stand up, the young child who piles wooden cubes on top of one another, the teenager who performs a high jump or dives into a pool are all unconsciously gathering a certain number of facts about gravity, although they do not see them as such. But you would be surprised how much one can learn by performing simple experiments and analysing them.

Let me start with a favourite of mine. To realize this experiment, you only need a large heavy book and a sheet of paper (of dimensions smaller than the book). Take the book in one hand, the paper in the other, and drop them simultaneously. The book falls like a stone, and the paper twirls around like a dead leaf. No wonder that Aristotle concluded that motion depends on the medium (air in this case). Let me now put the book on the paper, and drop them both. They reach the ground simultaneously, proving that the fall of objects is universal.

I know what you think: I was just cheating because I allowed the book to 'push' the sheet of paper. So let us do the opposite: I lay the paper on the book and drop them. Well, it is such an easy experiment that I encourage you to do it yourself before reading further, if possible in front of family, friends, or colleagues: success is guaranteed!

You will easily check that, again, book and paper fall in the same motion and reach the ground simultaneously. And, this time, you cannot tell me that the paper pushed the book. The reason is that the paper was protected by the book from friction forces in the air, and it thus fell as in vacuum, where all objects fall with the same motion.

This is a nice illustration of how one can uncover fundamental phenomena behind experiments simple enough to be analysed.

Another favourite experiment of mine illustrates the notion of inertia. It requires a table, a sheet of paper, and a glass full of water (and possibly a mop to clean up the mess). Place the paper along one side of the table, with 1 inch jutting out from the table. Place the glass on the paper (Figure I.1). Seize the free end of the paper and pull it out very suddenly. If everything goes well, the glass full of water stays on the table: it has more mass than the paper, and thus more inertia; it thus

Figure I.1 The experiment of the glass and the paper: inertia in reaction.

resists better to any sudden change of motion (in this case, the change from being at rest to moving sideways). If the experiment ended in a catastrophe (Grandma's last crystal glass shattered on the floor), it must be that the paper was too sticky (for example, because it was wet) or your pulling motion was not sudden enough. Try again with a kitchen glass.

You are now ready to reproduce Galileo's original study of the pendulum. You need a broomstick, a few metres of string, a 1-metre long board or stick, and various objects of different weights that can be easily tied to the string. Cut the string into strands of equal length (as many as the objects). Fasten them to the broomstick at equidistant locations, and tie the objects at the other end of the strings. If you now hold the broomstick horizontally, you have a series of pendulums that you can now set in motion (Figure I.2a). To initiate simultaneous motion, one may use the board to move them away from their equilibrium position in an identical way (Figure I.2b). Once set in motion, you will realize that the first few oscillations are identical, a sign of the universality of the motion of a pendulum (the objects have different masses). After a

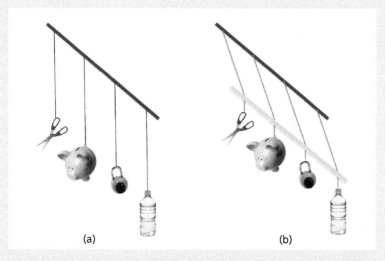

Figure I.2 Comparing pendulums with different mass but same length.

Figure I.3 Comparing pendulums with same mass but different length.
(Harvard Natural Sciences Lecture Demonstrations, available at
https://www.youtube.com/watch?v=yVkdfJ9PkRQ.)

while, friction in the air takes over and the oscillations are no longer synchronous.

You may now try to build a series of pendulums with the same mass but with different lengths. You will realize that they beat with different periods: the period thus depends on the length of the string but not on the mass or the nature of the object fastened to it (at least in vacuum). If the lengths of the strings are chosen to be linearly increasing with the position of the pendulum, you may see some beautiful patterns (Figure I.3), another manifestation of the universal character of the pendulum motion, of the universality of the laws of gravitation.

2

General Relativity:
From the Theory of Gravity
to a Theory of the Universe

We left Einstein in November 1915: at the time he was publishing his work on general relativity, summarized by equations that have since become known as Einstein's equations. These equations quantify the way space–time is locally curved as a function of a distribution of mass or more generally of energy. Thus, we are able to determine the geometry of space–time close to a massive star like the Sun and deduce the trajectory of light rays: they follow the shortest path, which, in the presence of curvature, is no longer a straight line. Einstein could have satisfied himself with these results; however, probably encouraged by the fact that gravitation governs the motion of all stars, he applied his equations to the whole Universe. This bold move marked the birth of modern cosmology.

But what was the known Universe in 1915? Not much besides our own Galaxy, the Milky Way. It will take yet years of observation and interpretation to understand that the Universe is, in fact, much larger and richer. But even though Einstein did not know about the marvellous worlds that expand beyond the Milky Way, the equations he derived still describe today our Universe in its largest dimensions. It is this story that I will retrace in this chapter.

Einstein's equations

What are Einstein's equations expressing? Very simply they relate the curvature of space–time to the energy content (in particular, but not only, in the form of mass since, following the famous relation $E = mc^2$, mass is energy) in any region of space–time.

Better than $E = mc^2$

According to legend, his editor having warned him that, for each equation that would appear in his book *A Brief History of Time*, the number of readers would be divided by two, Stephen Hawking is said to have contented himself with the iconic equation of special relativity, $E = mc^2$. I have not verified this story with him but I will risk writing Einstein's equations of general relativity:

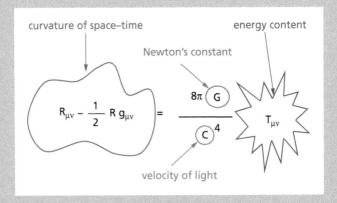

Let half of the readers content themselves to admire the aesthetic character of this beautiful equation and boldly move forward to the remainder of this chapter. For the others I will only point out that on this equation's left-hand side sits a quantity that characterizes the curvature of space–time and on the right a quantity that measures the energy content. Thus, the equation simply says that 'curvature = energy'. Two fundamental constants appear: the velocity of light noted c as was already familiar in the context of special relativity $(E=mc^2)$, and Newton's constant, noted G, which shows that this equation is indeed expressing a law of gravitation.

To better understand this, let us return to the phenomenon of curvature of light rays by the presence of a large mass.

When light draws space–time

In the absence of mass or more generally of any gravitational effect, light rays propagate in a straight line, i.e. along the shortest path. If light rays are parallel, we have known since Euclid that they will never meet. If now we let them pass very close to a massive star, they will be curved, all the more so the closer they graze the star. We can bet that, now deflected, they will meet further along their trajectory. We interpret this as an indication that space–time is curved by the star, at least in its vicinity.

To understand this, let us take a simple analogy. A sphere is curved: its curvature is simply determined by its radius. Take the familiar setup of the Earth. If, starting from the Equator and moving towards the North Pole, we follow two meridians of similar longitudes (Figure 2.1a), the trajectories close to the equator are parallel to one another and still they meet at the poles! It is precisely a consequence of the fact that the surface of the Earth/sphere is curved.

Let us now return to the free fall of material objects, more precisely by using the example of the falling lift in the previous chapter. We are at a certain point on Earth, say the Eiffel Tower, and we are floating in

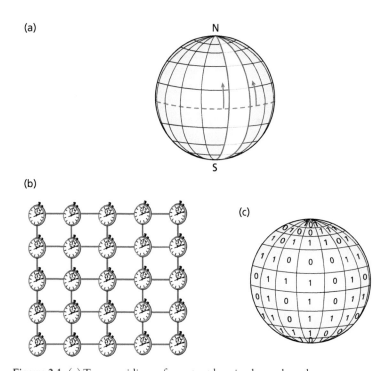

Figure 2.1 (a) Two meridians of constant longitude on the sphere.
(b) Space–time grid of flat space corresponding to a choice of metric.
(c) Space–time covering of the sphere where one uses locally at each point a grid of type (b).

the lift that is unhooked and is now in free fall. If we give a little push to the ball that is floating next to us, it will acquire a certain velocity and by inertia will keep it. Yet as seen from the exterior, it is just falling with an accelerated motion. But, in the reference frame of the lift, its motion is uniform. For this reason, this type of frame is called an *inertial reference frame*. The laws of classical physics are valid in this frame: in particular light rays travel in straight lines. We can check all this with a ruler and a clock that allow us to measure times and distances inside our lift/ frame, and thus turn local space–time into a grid. We describe this by saying that we have chosen a space–time metric: note that in 'metric' there is 'metre', hence 'measure'.

How is it then that light rays are propagating along curved lines? Let us take again our example of the Earth/sphere. The equivalent of the inertial frame described by Figure 2.1a would be a little sheet of engineering paper that we would put down at a given point, say corresponding to the location of the Eiffel Tower. What about light rays there? They follow locally the grid on the paper, but, because the Earth is curved, once they have travelled an infinitesimal distance they are on another point of the sphere where there is another piece of paper, and they follow the grid of this next piece of paper. In other words, they move to the next inertial reference frame. Of course, if these papers are very close by, they are almost identical and the effect is minute, but you will agree that, once they have been traveling from the equator to the pole the cumulative effect is important (Figure 2.1c): light rays which were parallel at the equator are now meeting at the pole.

We conclude that, if the metric (which allows us to make measurements at each point) is the same at all points, space is flat. If this metric varies at each point, as in the case of the sphere where we had to multiply the number of small papers, space is likely to be curved.

Indeed, curvature expresses the way the metric varies from one point to the next in space–time (in a way, the data given by all the sheets of paper in Figure 2.1c allow us to reconstruct the sphere and to measure its radius). Einstein's equations, which relate curvature and distribution in mass–energy, allow us to identify how this distribution of energy influences variations of the metric from one point to the next.

Note that in the example of Figure 2.1b we have mapped both space and time. The metric in general relativity is thus mapping space–time. This is important to note because we saw in Chapter 1 that space and time are intimately connected (dilatation of times, contraction of lengths in moving reference frames): it would be meaningless to treat space independently from time.

Following the effort of abstraction that has required this introduction to Einstein's equations, let us now return to Earth in the year 1915.

The Universe reduced to our Galaxy?

The Universe known at the time of the conception of general relativity was not so different from the one that William Herschel described in 1785 in his communication 'On the Construction of the Heavens' (Figure 2.2). It consisted of our own Galaxy, the Milky Way, an ensemble of stars that occupied fixed positions with respect to one another, and probably a large void around it. For example, no one had identified yet that the Andromeda nebula was outside our Galaxy.

To apply his equation to this *static* Universe, Einstein tried to find solutions that were independent of time. In order to do so, he had to modify his equations and introduce an extra term called the cosmological term, which depended on a new constant (besides Newton's constant) called the *cosmological constant*.

But more observations were being made, alongside estimations of astronomical distances. A debate raged in 1928 within the American National Academy. On one side was the astronomer Heber D. Curtis, who theorized that the Universe was made of many galaxies, some of

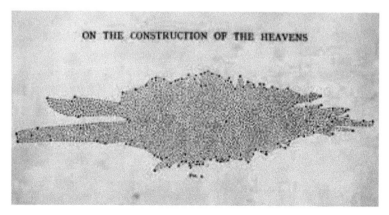

ON THE CONSTRUCTION OF THE HEAVENS

Figure 2.2 The Universe according to William Herschel in 1785.

which had already been identified in the form of spiral nebulae.[1] Opposing him was Harlow Shapley, who supported the view that the Universe was composed of a single large galaxy, whereas spiral nebulae were mere neighbouring gas clusters. This debate has become known in the history of astronomy as 'The Great Debate'.

In 1925 Edwin Hubble studied, with the aid of the Mount Wilson Observatory telescope, the Cepheids, which are variable stars in the Andromeda nebula M31. He showed that their distance from us is even larger than what Shapley thought to be the size of the Milky Way. M31 is thus its own galaxy, the Andromeda Galaxy, at a distance of 2.5 million light years away. This is an extragalactic object.

An expanding Universe

Immediately after the first elements of Einstein's theory on general relativity were published, solutions to his equations were proposed. As we have seen, a solution makes it possible to compute how, given a distribution of mass–energy, the metric varies as a function of space and time. As early as the end of 1915, which is one month after Einstein's publication, the German physicist Karl Schwarzschild identified how to solve these equations outside a spherical star. At the time Schwarzschild was in the army, stationed on the Russian front, so Einstein presented these results to the Prussian Academy during the first days of 1916, just before Schwarzschild died from a rare disease. Schwarzschild's solution was to play a central role in general relativity, in particular in the identification of black holes. We will encounter it again and again in this book.

Einstein had introduced the cosmological term in order for the solution to his equations to be time independent. However, in 1917 Willem de Sitter obtained a time-dependent solution of the same equations. Who was right? We can, in fact, show that the static solution initially proposed by Einstein suffered from instabilities. The de Sitter solution thus seemed to be a generic solution. It was promised a bright future since, as we will see, it is the basis of the theory of cosmic inflation, the period when the Universe's expansion is exponential. But if

---

[1] Interestingly enough, Immanuel Kant proposed this theory in his work *Universal Natural History and Theory of Heaven*, published in 1755, in which he defended the idea that nebulae were Universe islands too far away for us to identify individual stars.

time-dependent solutions existed, why should the Universe be static? In June 1922 the Russian physicist Alexander Friedmann published, to the great displeasure of Einstein, a theory of an expanding Universe.

This idea was soon confirmed by observation. In a first stage, the Belgian physicist Georges Lemaître, in an article published in French in 1927, then later Edwin Hubble in 1929, showed that extragalactic objects were moving away from us at a velocity proportional to their distance (Figure 2.3). The ratio between the receding velocity and the distance has become known as Hubble's constant.

How do we measure this recession velocity? By using what is called spectroscopic measurements, i.e. measurements of the wavelength of the light received from the stars. We know that the light emitted by objects has a characteristic colour, which means a characteristic wavelength. This wavelength is shifted if the object is in motion with respect to the observer who analyses the spectrum of the emitted light.

This is the famous Doppler–Fizeau effect that we have all experienced at one time or another, not with light waves but with sound waves: the sound produced by a fire truck is higher-pitched (the frequency is higher or equivalently the wavelength is smaller) when it approaches us and lower-pitched (the wavelength is larger) when it moves away from us

Figure 2.3 Georges Lemaître and Albert Einstein at the California Institute of Technology in 1933. © Bettmann/CORBIS.

(Figure 2.4a). Similarly, if we observe a galaxy that we know contains elements that emit light at characteristic wavelengths, these wavelengths will shift as the galaxy moves with respect to us. This spectral shift tells us whether the source is moving towards or away from us (resp. shifted towards the blue—or blue-shifted—or towards the red—or red-shifted; Figure 2.4b). It also gives a precise estimate of the velocity of the galaxy with respect to us. The name of Vesto Slipher, an American astronomer at the Lowell Observatory in Arizona, is associated with these first measurements.

By comparing the velocity identified with measurements of distance,[2] Lemaître and Hubble obtained what is now known as Hubble's law: the recession velocity of extragalactic objects is proportional to their distance. Figure 2.5 shows the results obtained by Hubble in 1929 and later in 1931; distances are in megaparsecs, Mpc, which correspond to 3.26 million light years. The size of our own Galaxy being on the order of one hundredth of a megaparsec, the objects considered are indeed extragalactic.

The proportionality factor is called, as we have said, Hubble's constant. Because its value depends on measurements of astronomical distances which are estimated step by step from the closest astrophysical objects to the most distant, and are thus prone to multiple errors (we will come back to this in Chapter 6), the determination of its exact value has fluctuated a lot over the years, before recently stabilizing around 70 km/s/Mpc.

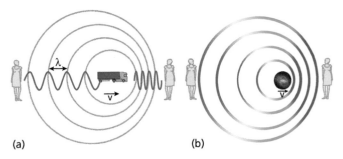

(a)   (b)

Figure 2.4 The Doppler–Fizeau effect with sound (a, truck) and light (b, galaxy). In the two cases the source is moving from left to right.

---

[2] I should mention here Henrietta Swan Leavitt, whose pioneering work in the measurement of extragalactic distances was long forgotten before being rediscovered.

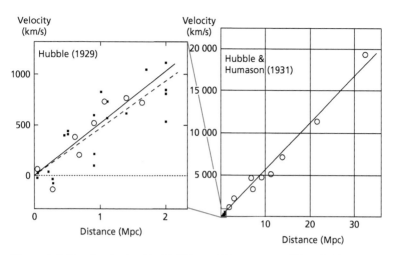

Figure 2.5 Results obtained by Hubble and Humason in 1929 and 1931 showing how the recession velocity of extragalactic objects varies linearly with their distance from us.

How do we interpret this result in light of what we know? First, we must consider the Universe as an entity that has its own dynamics, summarized by Einstein's equations. To state that the Universe is expanding amounts to saying that the distances mechanically increase because the metric depends on time (a little like a ruler increasing in size with time). Thus, other galaxies are not receding from us because they are moving with respect to us, but because they are located at a given point and it is the Universe's texture itself that is dilating. It is in this sense that we must understand the analogy of an inflating balloon (Figure 2.6): the individual galaxies are points on the balloon; they automatically recede from one another when the balloon inflates. This example shows that there is no specific point of view: wherever you are on the surface of the balloon, you see the other points receding away.

A question may come to mind: in the case of the Universe, who is inflating it? Or in other words, what source is providing the energy necessary to expand the Universe? The question is legitimate and shows that you have mastered well the law of conservation of energy.

Unfortunately, it is difficult to answer in the framework of general relativity because there is no universal definition of the energy of a system like the Universe. Is it a limitation of the theory to which we should

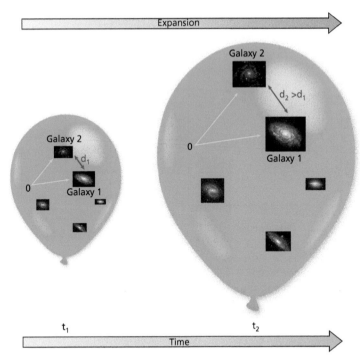

Figure 2.6 The expansion of the Universe compared to the expansion of an inflating balloon. The distance between two points (Galaxies 1 and 2) mechanically increases with the expansion of the balloon.

find a remedy, or should the law of conservation of energy be replaced by a more complex law? You see, we have not yet reached the end of the second chapter and already here is a question without an answer. If you expected to have answers to all your questions, you might be disappointed. However, the field is thriving, and in our odyssey we will encounter other questions that remain unsettled. It already shows significant progress to be able to formulate some of these questions more precisely.

Since we are dealing with delicate questions, here is another one: according to the Hubble law the further an object is, the faster it recedes from us. Are there objects then that recede from us faster than the velocity of light (one talks of superluminal velocity)? You will find in scientific books all sorts of answers to this rather straightforward question. The answer turns out to be 'yes', which might surprise you since the velocity of light in a vacuum is the maximum value of the velocity of a material body. But the recession velocity of galaxies is not, in the strict sense, a physical velocity. Let us recall that these galaxies are not moving: it is the space between them and us that is undergoing dilatation.

A couple of last questions that it is important to clarify before moving forward: are all objects in the Universe undergoing this expansion, and can one measure the effect of this expansion in the laboratory? The answer to both is no. Galaxies and stars were formed by gravitational accretion of matter: matter clumps together, and the gravitational attraction of the lumps counteracts the effect of the dilatation of the Universe's texture. If you have diluted flour into milk, you know that once lumps form, adding more milk will not break them up. We say that the matter that formed the burgeoning galaxy decoupled from the universal expansion by falling into the 'gravitational well' of the original lump. On the other hand, of course, the different gravitational wells that are the galaxies are receding from one another because of the expansion of the Universe. This is why we first had to identify the extra-galactic objects before becoming conscious of the expansion and measuring it. The stars of the Milky Way, our own gravitational well, remain at fixed distances from our Solar System. This explains why in 1915 Einstein could not imagine anything but a static Universe.

Since then, one has even discovered that the expansion of the Universe itself is dynamical, i.e. time-dependent. In fact, the constant of the Hubble law, which we called the Hubble constant, does not have the same value for the relatively near galactic objects studied by Hubble, or for much more distant objects. The Hubble constant has a history! This is why we refer to it as the *Hubble parameter*, keeping the expression *Hubble constant* for the most recent value of this parameter: it measures the expansion rate of the Universe *today*.

The Big Bang model

Identifying the expansion of the Universe has make it possible to develop a coherent model of its evolution, or cosmological model, which I am now going to present.

First of all, even if the most striking feature when we observe the Universe is the presence of structures such as galaxies or galaxy clusters, the Universe in its largest dimensions is relatively homogeneous. This means that if we look at a region of the sky sufficiently large to contain a large number of galaxy clusters, its properties are very similar to those of another region of similar size. If we define the *energy density* as the ratio of energy contained in this fraction of the sky (in the form of mass, radiation, etc.) to its volume, then, on average, the energy density does not depend on the region considered: it characterizes what we call the average energy density of the Universe.

You may think that I give the observer an exceedingly important role: wouldn't an observer localized elsewhere in the Universe measure a different energy density? This would be true if I paid attention to small distance scales, i.e. what is close to the observer. For example, we are in the Milky Way; an observer outside this Galaxy would probably see a very different sky. However, we are interested in the largest dimensions of the Universe, for which this type of 'galactic' detail does not play any role. And we hypothesize reasonably, if the Universe is homogeneous, that we are not placed at a specific point in the Universe: any other observer, located at a different point, would identify very similar average properties of the Universe at large, and would in particular measure the same average energy density.

However, because the Universe is expanding, a quantity of energy distributed over a certain volume will be diluted into a increasingly larger volume as time goes on. We conclude that, through the evolution of the Universe, energy density has decreased.

A consequence of Einstein's equations is that the rate of expansion, which is measured by the Hubble parameter, varies in the same way as the total energy density. Thus, the rate of expansion decreases with time as well.

Another important concept is *temperature*. The temperature of a material body is directly related with the agitation of its molecules. For example, the greater the temperature of a gas in a closed box, the greater the energy (kinetic) of its colliding molecules. If we decrease the size of the box, this agitation increases: temperature rises. If, on the other hand, we increase the size without pumping energy into the gas, temperature drops. This is exactly what happens in the Universe. We can liken the Universe to a large 'box' where particles, atoms, molecules are in thermal agitation due to numerous collisions. We can define a temperature corresponding to this thermal agitation. But, because the Universe is expanding, the box size increases without energy being added to the gas's particles. The average temperature thus drops with time.

Because the temperature of the Universe is directly related with the thermal agitation of molecules, physicists prefer to measure this temperature with respect to absolute zero, i.e. −273°C, where this agitation would be totally frozen. We thus use Kelvin, noted K, shifted by 273 units with respect to the Celsius degree. Thus, 0 K = −273°C, 273 K = 0°C and, for instance, 3000 K corresponds to 3000 − 273 = 2727°C.

This leads thus to a model of an expanding Universe that, in the course of its evolution, becomes less and less dense, and correspondingly colder. Conversely, if we imagine going back in time, this Universe gets more and more dense and warm until an early time where it reaches an infinite density and temperature. We then speak of a singular behavior, or of a *singularity*. Is this signalling that general relativity and quantum physics must be replaced by a new theory that encompasses both of them? Or is it a sign that the physical description of nature has reached a fundamental limitation? We will return to this question in Focus IV.

This initial singularity that appears in our model of expanding Universe was scoffed at by the British physicist Fred Hoyle, sworn opponent of the model, in a 1949 BBC radio show; he tried to ridicule it by describing it as a 'Big Bang', a picturesque term that remained. We thus speak of the Big Bang model to denote this model of an expanding Universe that becomes less dense and colder with time. One also refers to the Big Bang to denote the initial singularity that appears in this model. Beware of not confusing the two notions: there is much more to the Big Bang model than the initial Big Bang; in fact, it may well describe the full evolution of the Universe, except for the Big Bang singularity. This might be the posthumous revenge of Fred Hoyle!

History of light

We have all learned that light has a finite velocity. Galileo unsuccessfully tried to measure this velocity, which was then estimated by the Dane Ole Christensen Romer in the seventeenth century using astronomical observations of the eclipse cycles of Jupiter's satellite Io (and their time delay following the position of Jupiter with respect to the Earth).

For a precise measurement, we have to wait till the nineteenth century for the competition between Hippolyte Fizeau and Léon Foucault in Paris. As a matter of fact, the interest was focused not so much on a precise evaluation than on the nature of light, corpuscular as defended by Newton or wave-like as proposed by Christiaan Huyghens. In order to explain the phenomenon of refraction, Newton had assumed that the velocity of light increased in a denser medium because the gravitational pull was greater. According to the wave theory, on the contrary, the velocity of light is lower in denser media.

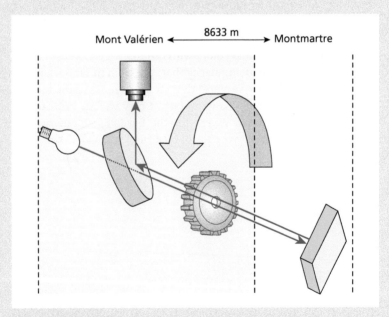

Figure II.1 The dented wheel setup used by Fizeau in 1849 to measure the speed of light.

In his setup mounted on the hill of Montmartre in the centre of Paris, Fizeau used a dented wheel that rotated at high speed. Light propagated transversely to the wheel, grazing it in such a way that it passed through one of the dents. It then travelled a large distance (8.6 km) to the suburb of Mont Valérien where it was reflected back, returning to the setup, and passing through another of the wheel dents, thanks to a careful tuning of the wheel velocity (Figure II.1). From the number of dents in-between and the rotational velocity, Fizeau obtained the velocity of light.

Foucault, on the other hand, used a rotating mirror. The principle was rather similar. Light was reflected off a mirror rotating at high speed to be sent to a fixed spherical mirror: during the time it took for the light to return to the moving mirror, it had been rotated by a certain angle. The light ray was then reflected back to the original source, but with a deflection that was twice this angle: the measure of this angle gave the travel time if the rotation velocity was known (Figure II.2).

Foucault won the competition by a mere six weeks, in April 1850: light propagates in the air faster than in water. Light is thus a wave. In the next century, quantum mechanics (Louis de Broglie) and Einstein would reconcile both sides of the debate: light is both a wave and a particle (the photon), and its maximal velocity is reached in the vacuum.

To obtain a more precise value of the velocity of light, Foucault conducted a new experiment at the Paris Observatory in 1862: he measured a value of 298,000 km/s, very close to the exact value in a vacuum, $c = 299{,}792{,}458$ m/s.

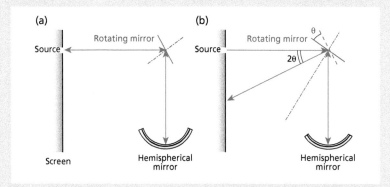

Figure II.2 Foucault's experiment to measure the speed of light: (a) initial position and (b) after a rotation of the mirror at angle θ.

I am using the word 'exact' because the Seventeenth General Conference on Weights and Measures decided in 1983 to define the metre as the distance covered in a vacuum by light in a time of $1/299{,}792{,}458$ s. The old prototype metre bar kept at the Bureau International des Poids et Mesures in Sèvres, near Paris, was abandoned and the metre defined in such a way as to fix once and for all the fundamental constant that is the velocity of light, c, to the value 299,792,458 m/s.

The fact that the velocity of light is finite has important consequences for observation of the Universe. Indeed, the further we look, the earlier in time we see, which is why in astronomy the unit *light year*, i.e. the distance light travels in one year (31,557,000 seconds), is often used. We can easily check that 1 light year amounts to 9,460 billion km, or, using powers of 10, 9.4×10^{15} m. Since we are talking about units, let me add that you will often encounter in this book (as we already have at the start of this chapter), and in most astronomy books, the parsec (pc), a unit of distance that corresponds to 3.26 light years, and the megaparsec (Mpc), 1 million parsecs.

The sky that we observe is a formidable time machine: in principle, if we could look at distances of 14 billion light years (corresponding to the

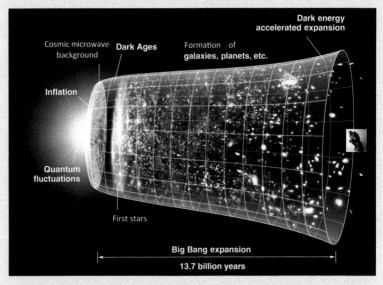

Figure II.3 The history of the Universe represented along a time axis.

age of the Universe), we could see the Big Bang! We will see in Focus VII that it is a little more complex, but this underlines the power of observation: *we are observing in space and time*. But have we fully apprehended the meaning of this statement?

For example, what do we know of the present state of a galaxy that is 2 million light years away? Nothing really! It may have suffered a catastrophic fate while the light emitted 2 million years ago has been travelling towards us. How can we get a knowledge of its state today? We must wait another 2 million years (in fact, a little more because of the expansion of the Universe). And can we get information from its past,

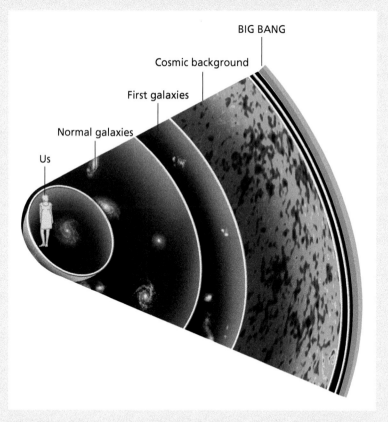

Figure II.4 The history of the Universe from the point of view of an observer.

beyond two million years ago? Not directly: light emitted, for example, 3 million years ago reached our Earth 1 million years ago.

As a matter of fact, I should qualify the statement made earlier: the observation of light emitted by distant astrophysical objects (and this is true as well of other cosmic messengers, such as cosmic rays or neutrinos) does not allow us to observe in space and time; rather, it provides us with a cross section in space–time.

We will come back to this in Focus III. Let me just point out here that there are two complementary representations of space–time. One (Figure II.3) describes the Universe along a time axis that starts at the Big Bang to end today, or even in the future. The part of the Universe observable today (which is presumably not the whole Universe) is reduced to a tiny spatial region, almost a dot, at the time of the Big Bang.

The other one (Figure II.4) is a spatial representation centred around the observer (us) today. It is formed by a series of concentric shells, each of which corresponds to a given distance with respect to the observer, and thus to a given period in the evolution of the Universe. The largest (furthest) shell represents the Big Bang.

The two representations are equally valid and they give a complementary understanding of reality. We will now focus on the details of the Universe's evolution as it is understood today, and will alternate between the two representations.

3

Observing the Universe

Natürlich, wenn ein Gott sich erst sechs Tage plagt,
Und selbst am Ende bravo sagt,
Da muss es was Gescheites werden.

(Of course, when for six days a God his labour plies,
And, of his own accord, then 'Bravo' cries,
We may expect good workmanship to see.)

JOHANN WOLFGANG VON GOETHE, *Faust* (1806)

Now that we have a more precise idea of how we may grasp the history of the Universe, we will continue exploring it, making use of not just our eyes but the most modern observation tools. The further we observe, the more we observe in the past. These modern instruments will allow us to get back to a primordial era in the Universe's evolution, 380,000 years after the Big Bang to be precise.

We will use two specific ways to bookmark the different periods that we will encounter as we observe more and more distant astrophysical objects. One is the number of light years, which is the distance the light emitted by an object must travel to reach us: it allows us to identify the actual time at which we observe the object. Since the Andromeda Galaxy is 2.5 million light years away, its light takes 2.5 million years to reach us; thus, we observe this galaxy as it was 2.5 million years ago.

The other one is the spectral redshift. We encountered it in Chapter 2: the spectral redshift, traditionally denoted by the letter z, measures the shift in frequency (or equivalently in wavelength) of light due to the recession velocity of the extragalactic object that emits it.

There is a complementary way to understand the spectral redshift. The wavelength of light emitted by an element in a distant galaxy could be used by its inhabitants to define a unit length. We can use the light received on Earth to define our own unit length. The ratio between the two is equal to 1 plus the spectral redshift, i.e. $(1 + z)$, which measures the dilatation the Universe has undergone since this light was emitted.

Spectral redshift

Since galaxies appear to move away from us, the wavelength observed appears to be larger than that of the light they emit. The ratio between the two is thus larger than 1: it is equal to 1 plus a positive quantity, noted z, which is called the spectral redshift.

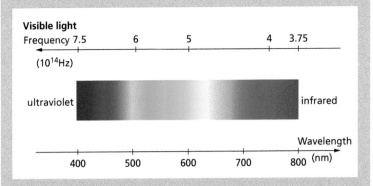

The wavelength of visible light varies between 390 nanometres for violet and 780 nanometres for red light (1 nanometre is 1 billionth of a metre). Since the spectral redshift is positive, it shifts wavelengths towards larger values, i.e. towards the red: hence the name *red*shift. Actually, the light emitted by distant galaxies is shifted so much that it is observed on Earth beyond 780 nanometres, i.e. in the infrared domain.

This phenomenon applies, in fact, to waves all over the electromagnetic spectrum, not just to the visible light (and indeed to all waves).

Thus, the region of the Universe accessible to our observations today, what we call the observable Universe, had a size 1001 times (i.e. about a thousand times) smaller at a time when the redshift was $z = 1000$. This explains why it is very practical to use redshifts to characterize a past era in the Universe's evolution.

A last question to check that you are mastering well the notion of redshift: what is the redshift today on Earth? Well, 0, since the ratio between the wavelength of light emitted (today) and light observed (today) is obviously 1, i.e. $1 + 0$. Our redshift clock marks 0 today and here on Earth and runs to higher redshifts as we go back in time, further away in distances.

Heading for the past

Let us observe more distant objects, hence towards more and more past events. The celestial objects closest to us are our star, the Sun, and its planets. Light takes, for instance, 40 minutes to travel from Jupiter to Earth. Beyond Saturn, Uranus, and Neptune, the Kuiper belt is made of objects composed largely of ice. Further, the Oort cloud, thought to be the remnant of the original proto-planetary disc, is probably the source of all long-period comets. Its external frontier forms the boundary of the gravitational pull of the Solar System, at 1 to 2 light years from the Sun. It is approximately one quarter of the distance to the nearest star from the Sun, Proxima Centauri of the Alpha Centauri system (at 4.37 light years). Before leaving our system once and for all, let us note that we should not confuse the time at which we observe astrophysical objects and their age. This is obvious in the case of the Solar System, which is believed to have formed some 4.5 billion years ago.

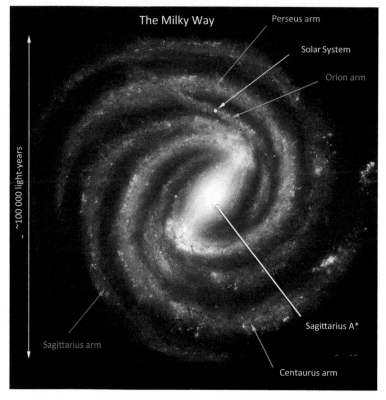

Figure 3.1 Our Galaxy, the Milky Way, with some of its spiral arms, the Solar System on the Orion arm, and the galactic centre, Sagittarius A*. © R. Hurt (SSC), JPL-Caltech, NASA.

As we move further away, we discover the stars of our Galaxy, the Milky Way, which contains a few hundred billions of them. It is a galaxy of the spiral type, shaped as a disc of more than 100,000 light years in diameter and 1000 light years in thickness, with a flattened bulge at the centre (Figure 3.1). The Sun is located somewhat on the side, at about 26,000 light years from the centre, on one of the spiral arms (Orion). It is close to the equatorial plane (5 light years).

The stars of the Galaxy rotate around its centre, with a velocity that depends on their location. This velocity increases from the centre and then stabilizes, away from the bulge, at values between 210 and 240 km/s.

The galactic centre hosts a compact object of very large mass, Sagittarius A*, that is thought to be a black hole with a mass equivalent to 4 million times the mass of the Sun. Physcists nowadays think that most galaxies have a central black hole of a mass between hundred thousands and tens of millions of solar masses (we will return in detail to these massive black holes in Chapter 7). Sagittarius A* and its colleagues will be playing an important role in what follows (Chapters 7 and 10).

Once we have passed the last stars of our galaxy (that are at a distance of 78,000 light years), we encounter a few dwarf galaxies that can be considered as satellites of our own. The best known are the Large and the Small Magellanic Clouds, respectively at 179,000 and 210,000 light years.

The nearest galaxy is Andromeda, at about 2.5 million light years away. Andromeda is also a spiral galaxy, expected to contain some 1000 billion stars (Focus III). With the Milky Way and the small galaxies around, it forms what is called the *Local Group* (Figure 3.2).

Andromeda is the largest of the Local Group. Its more luminous central part is visible with the naked eye on a moonless night. The Local Group is itself part of a larger structure, or galaxy cluster: the Virgo cluster. Our intergalactic address is, thus, Earth, Sun, Milky Way, Local Group, Virgo cluster: a piece of information that may become precious when we want to find our way back on this journey through space and time.[1]

We know that galactic objects recede from us because of the expanding Universe. To this recession velocity is added a local velocity, called

[1] An illustration of the way our mental images betray us: we are not travelling, but stuck on Earth observing the night sky.

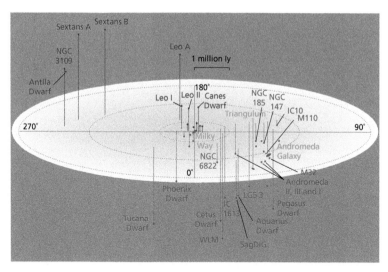

Figure 3.2 The Local Group to which the Milky Way and Andromeda galaxies belong.

peculiar velocity, which results from the gravitational attraction within the cluster. In the case of Andromeda, the peculiar velocity within the Local Group is sufficient for Andromeda to be moving closer to us at a velocity of about 110 km/s: the light received from Andromeda is thus blue-shifted (and *not* red-shifted). A collision is expected in 4 billion years, which leaves us time to prepare. We will see that such galaxy collisions play a fundamental role in the structuration of matter in the Universe. The galaxy that would result would probably be of elliptic type, that is, with the form of an ellipsoid.

Let us keep moving away. We will encounter many galaxies. The picture of Figure 3.3, taken by the Hubble telescope, shows about ten thousands of them (out of a total of a few hundred billion that the observable Universe contains). We can identify galaxies with a recognizable shape, of spiral or elliptic type, but also galaxies of a more irregular shape, that are more ancient. These older galaxies are moving faster away from us and the light they emit should be shifted towards the red or infrared. But they appear bluish in the picture. This is because they are home to active zones of stellar formation, a formation that emits ultraviolet light detected by Hubble (and appears bluish in the picture). Such zones appear in galaxies that are at a distance between 5 and 10

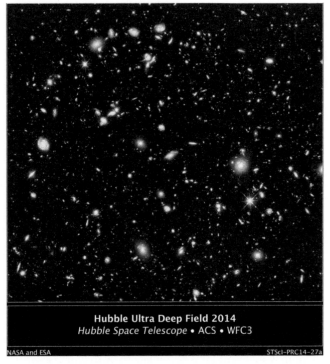

Hubble Ultra Deep Field 2014
Hubble Space Telescope • ACS • WFC3

NASA and ESA   STScI−PRC14−27a

Figure 3.3 Image taken from the so-called deep field survey of the Hubble telescope. © NASA, ESA, R. Ellis (Caltech), and the UDF 2012 Team.

billion light years away. The most ancient galaxies identified in the image are seen a few hundred million years after the Big Bang! By moving earlier in time, one reaches the dark ages, where stars have not yet formed to illuminate the Universe.

The first galaxies had an irregular shape. One thinks that it is through a succession of galaxy collisions and mergers that the large regular, elliptic or spiral, galaxies that we know were formed. This is why such galaxies are much more recent. One can get an idea of such galaxy mergers by searching through the billions of observable galaxies; the double systems present the characteristics of a collision at different stages of the merging process. Figure 3.4 gives some examples, extracted from the rich data bank of the Hubble telescope.

These first galaxies were also, as we said, the home of an active formation of stars, up to hundreds or thousands per year (to compare

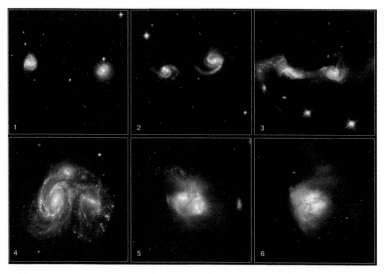

Figure 3.4 Several images taken by the Hubble telescope showing the different stages of the collision and merger of two galaxies (each picture corresponds to a different binary system). © NASA, ESA, the Hubble Heritage Team.

with a rate of one or two for nearby galaxies). The space mission Herschel has recently identified that this activity is not as important as expected: there seems to be a deficit of gas in these galaxies, probably due to their central black hole's activity, which would expel part of the gas outside the galaxy. It may surprise you to see a black hole expelling anything when you thought that everything falls into a black hole. We will come back to this in Chapter 7.

There is finally one key element in the dynamics of galaxies that we have not yet introduced: *dark matter*.

Galaxies, galaxy clusters, and dark matter

Dark matter is another
Matter. Cosmologists don't know.
The physicists do not.
The stars are not.

FREDERICK SEIDEL, *Invisible*
Dark Matter, The Cosmos Trilogy (2003)

Dark matter was identified as early as 1933 by Fritz Zwicky, who was studying the Coma cluster, a large cluster of more than 1000 identified galaxies that lies 320 million light years away. More precisely, Zwicky was studying the distribution of the galaxy velocities in this cluster. By studying their dynamics using Newton's law, he could infer the mass distribution: the mass was some 400 times larger than that expected by measuring the luminosity of these galaxies (the ratio mass/luminosity is typically for galaxies a factor two to ten times larger than the ratio measured for the Sun). There must have been some nonluminous form of matter in this cluster or within its galaxies. This was largely ignored and then gradually rediscovered and systematically studied throughout the years 1960 to 1970, in particular by the American astronomer Vera Rubin.

This exploration is based on the motion of stars within the galaxies. These stars, such as our Sun, which are mostly gathered into a flat structure shaped as a disc, are rotating around the centre of the galaxy. A simple calculation using Newton's gravitational force (no need to use relativity here!) shows that their rotational velocity depends on their distance to the centre. The further away from the centre one goes, the larger the rotational velocity, at least until one reaches the disc rim: the velocity of the more distant stars is then expected to decrease with distance. But observation of many galaxies shows that this velocity remains constant very far from the galaxy centre, or from the disc (see Figure 3.5).

To explain these curves, called rotation curves, we believe today that non-luminous matter is present in the galaxy at larger distances from the centre than luminous matter is. We consider today that each galaxy has a quasi-spherical halo of dark matter. This dark matter is also present in galaxy clusters, and in the whole Universe. Its nature remains, for the time being, unknown, even though particle physics provides many potential candidates.

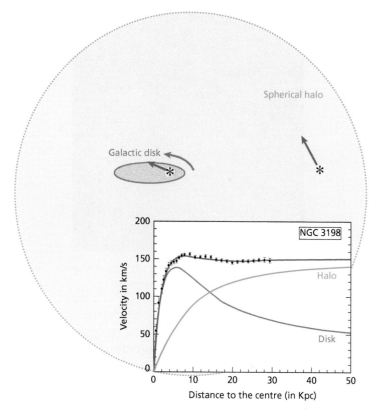

Figure 3.5 Rotation curves for the spiral galaxy NGC3198 (at about 40 light years from us) given the velocity of stars as a function of their distance to the centre (in kiloparsec (kpc), i.e. 3200 light years). The curve in blue (noted disk) corresponds to ordinary matter in the galactic disk; the curve in green (noted halo) corresponds to matter in a spherical halo. The total (in red) of the two curves is in good agreement with observation (in black).

The presence of dark matter is also essential for understanding the dynamics of how galaxies form. Numerical simulations show that first dark matter gathers gravitationally around local concentrations of matter, kind of like gravitational seeds, the origin of which we will elucidate in Chapter 5. Dark matter eventually forms large structures linked by filaments, surrounding emptier regions or 'voids' (Figure 3.6).

The primordial gas, which is ordinary matter formed of protons, neutrons, and electrons, then falls into these concentrations of dark

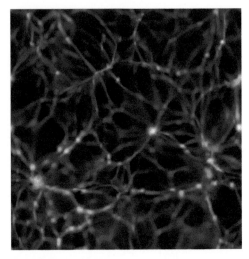

Figure 3.6 Distribution of dark matter in the early Universe as obtained by numerical simulations. © V.Springel / Virgo Consortium.

matter (that act like gravitational wells) to form the galaxies and galaxy clusters that we observe. Structures shaped like a sheet, such as the 'Great Wall', a concentration of some 500 million by 200 million light years with a thickness of only 15 million, which could correspond to the filaments surrounding the voids, have been discovered.

We can understand how matter gets progressively structured as the Universe evolves. Dark matter gravitationally falls into local concentrations of matter, a phenomenon that accelerates as time passes. Thanks to gravitational attraction again, ordinary matter is pulled into these gravitational wells and forms the galaxies that we see. These galaxies are first irregular and not very massive. By a process of recurring collisions/ mergers, they become more massive and regular until they become the large regular massive galaxies that we encounter in our Local Group. We can see here the amazing structuring power of gravity! And this process continues with the appearance of planetary systems, of life, and finally of the human brain, which may represent the ultimate stage in development and complexity. Which part did gravity have in this?

Conversely, as we move back in time, structures become more irregular and eventually crumble. This coincides with a process where the Universe becomes increasingly homogeneous. Because, while going back into the past, the average temperature increases, even microscopic structures such as molecules and atoms break into pieces. At a certain stage, thermal fluctuations are energetic enough to break atomic bonds and free elementary particles. The primordial (meaning 'very early') Universe is then a soup of elementary particles. It is thus a remarkable laboratory for fundamental physics. We will come back to this aspect in the next chapter. For the moment, let us focus on a key stage on the road back to the primordial Universe, which is known as the *recombination era*.

Recombination and the cosmological background

When the spectral redshift was 1100, i.e. when the observable Universe was 1101 times smaller, we are at approximately 380,000 years after the Big Bang. The Universe is then very hot: the temperature is 3000 K. The energy of thermal fluctuations is sufficient to break the bounds between electrons and nuclei that form the atoms.

Let us take the example of the hydrogen atom: it is the simplest atom, but also the most abundant one in this primordial era. It consists of an electron of negative charge ($-e$) and of a proton of positive charge ($+e$). The hydrogen atom is thus electrically neutral ($-e + e = 0$). Beyond 3000 K in the primordial Universe, the electron is torn away from the proton: we say that hydrogen is ionized (Figure 3.7).

Light is an electromagnetic wave: it is emitted and absorbed by electric charges. It thus interacts with electrically charged bodies but not

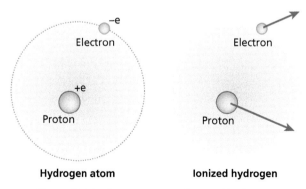

Hydrogen atom   **Ionized hydrogen**

Figure 3.7 Hydrogen atom and ionized hydrogen.

with neutral ones. Thus, light that propagates through the Universe ignores hydrogen atoms (neutral) but interacts with an electron or a proton, and can even be trapped by them if there is a sufficient amount of them (remember that the primordial Universe is dense).

This is why the Universe today is transparent to light. Apart from the dense astrophysical objects, it is occupied by neutral hydrogen clouds that do not catch light.

On the other hand, if we go back to times prior to the period when the temperature is 3000 K, then hydrogen, and more generally any matter, is ionized: the Universe is filled with what is called an ionized plasma, opaque to light. This means that any light emitted is immediately reabsorbed: light cannot propagate in such a medium.

I take this opportunity to clarify a misconception that may cause misunderstanding when we discuss an opaque primordial Universe. The Big Bang is often represented as a flare of light. This is just a picturesque way of showing that some energy was released, but we must caution that, if there ever were an emission of light, it would have been immediately absorbed.

What happens when temperature drops below 3000 K? The thermal movement of particles becomes insufficient in preventing electrons and protons from combining and forming *neutral* hydrogen atoms. This happens throughout the Universe, which becomes filled with neutral hydrogen. Light is no longer trapped by charges: it travels freely. *The Universe becomes transparent.* This period is called recombination (of hydrogen): more properly, we should say 'combination' since hydrogen atoms were never formed previously.

What does this all mean for us observers? When we look sufficiently far away to reach this period in time, our vision is stopped by an opaque wall that limits the dark region where light cannot travel: the primordial Universe appears to us as a black body. Does it mean that we see nothing? Not really, because a black body emits radiation!

To understand this, we must return to the birth of quantum mechanics at the beginning of the twentieth century, 1900 precisely, with Max Planck. You probably know that the colour of an object is related to the frequency of the electromagnetic radiation (light) reflected by its body: a rose is red because it absorbs all light frequencies except the red one that it reflects. Ideally, a black body is an object that reflects all electromagnetic radiation (whatever its frequency). But you know that a *black* piece of charcoal emits red light when heated, possibly white light if heated at an even higher temperature: we say that it becomes incandescent. It had indeed been noticed at the end of the twentieth century that a black body heated at a given temperature emitted a radiation whose spectrum was characteristic of the body temperature. The physical reason was known (thermal agitation of the body's atoms), but the shape of the spectrum was not understood. It was Max Planck who finally found the key: the emission of radiation is not continuous but is realized through grains of energy or *quanta*. Each grain of light carries an energy proportional to the frequency of its light, the proportionality constant being a new fundamental constant, called from then on Planck's constant, and noted h.

The mechanics of quanta, or *quantum mechanics*, was born, even though, in the beginning, Planck had difficulties digesting it and considered this was only a mathematical artefact. It was Einstein who understood the corpuscular nature of light: the grain of light is the photon. *Light is both a wave and a particle*. It reveals aspects of its dual nature depending on the type of physical process. The opposing views of Newton and Huyghens were finally unified!

Let us return to the opaque wall that our 'eye' encounters when it observes far enough to reach a time when the Universe was only 380,000 years past the Big Bang, i.e. the epoch of recombination (the corresponding redshift is 1100). It is a perfect black body heated at 3000 K (the temperature of the Universe then) and we should observe the corresponding radiation. This radiation was indeed predicted in the 1940s by George Gamow, and discovered by chance in 1964 by Arno Penzias and Robert Wilson (who received the Nobel Prize in 1978). Penzias and Wilson were working on a new model of antenna for Bell Labs when they discovered a background radiation that they tried to characterize. It was eventually the homogeneous and isotropic (i.e. identical in all directions) character of this radiation that convinced the scientific community of its cosmic nature: the sources of this radiation appeared to be evenly distributed all over the sky.

This electromagnetic radiation falls in the microwave frequency domain (between infrared and radio waves). This corresponds precisely to the radiation emitted by a black body heated at 3000 K and having been red-shifted by a factor 1100: the associated temperature is then divided by the redshift factor, i.e. $3000/1100 = 2.73$ K. This was spectacularly confirmed by the COBE satellite in 1990: on board, the FIRAS instrument, supervised by John Mather, was devised to measure precisely this radiation and to compare it with an artificial black body embarked on the mission. The agreement is impressive (Figure 3.8).

This radiation is called cosmic microwave background, often abbreviated as CMB. The corresponding photons were produced 380,000 years after the Big Bang and propagated to us unperturbed, since from the recombination onwards, the Universe has been transparent. This is why we talk of the first light: previously, the Universe had been opaque to photons; every light emitted was immediately trapped. Note that we talk here not just of visible light, but of all the light spectrum, all the electromagnetic spectrum.

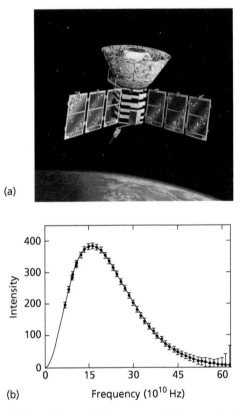

(a)

(b)   Frequency (10^{10} Hz)

Figure 3.8 (a) COBE satellite and (b) spectrum of the cosmological background radiation (intensity of radiation versus frequency, the bars representing the observational data) compared to the spectrum of an ideal black body at temperature 2.735 K. (a) © NASA.

We can only marvel at the fact that this black-body radiation, the understanding of which marked the birth of quantum mechanics and of the laws of the microscopic world, appears as well at the level of the largest distances in the Universe. This mysterious link between the inner space of the particle world and outer cosmic space will be a recurring theme throughout this book.

The COBE satellite brought another exceptional result: the instrument DMR under the responsibility of George Smoot discovered anisotropies within the CMB radiation. Not large enough to question its cosmic origin, the equivalent black-body temperature of the radiation had variations at the level of 1 to 100,000 (or if you prefer some 10^{-5} K), depending on the direction one was observing. But this is another story that we will examine in Chapter 5.

John Mather and George Smoot received the Nobel Prize in Physics in 2006.

FOCUS III
The Universe as seen from the Andromeda Galaxy

今は昔 (ima wa mukashi)
This is now a thing of the past
今昔物語集, *KONJAKU MONOGATARI SHŪ*
(Formula introducing each of the 1,059 stories)

Let us imagine that we are on a planet orbiting a star in the Andromeda Galaxy. What would we Andromedians see of the Universe?

Let me start by presenting this Andromeda galaxy, also called Messier 31 (M31) or NGC224 (Figure III.1). It is a galaxy of the spiral type, with an approximate diameter of 140,000 light years, that contains approximately 1000 billion stars. It is thus larger than the Milky Way, even though its total mass of about 1000 billion solar masses is slightly smaller than the mass of the Milky Way. The centre of the galaxy hosts a star cluster with a dual structure: one of them includes a massive black hole

Figure III.1 The Andromeda Galaxy and its two satellite galaxies, M32 and M110 (respectively at 3.00 and 11.00 o'clock from the centre of Andromeda). © Niedzwiadek78.

of about 100 million solar masses! The galactic disc is warped, probably under the effect of satellite galaxies M32 and M110.

Infrared images from the space telescope ISO of the European Space Agency have shown the presence of ring-shaped accumulations of dust and gas, some of them as close as 30,000 light years from the centre. The centre of this structure is shifted from the galactic centre. This could be due to the head-on collision of Andromeda with the small elliptic galaxy M32 (Figure III.2) 210 million years ago: the collision would have stripped a large fraction of mass from the smaller galaxy.

If we observe the Universe from the Andromeda Galaxy, we first see in the sky stars of the galaxy and identify extragalactic objects, first the satellite galaxies M32 and M110, then other objects of the Local Group, in particular at a distance of 2.5 light years the Milky Way or Via Lactea Galaxy, a spiral galaxy rather similar to Andromeda.

If we were searching for evidence of life in this galaxy, we might identify perhaps, on the Orion arm (Figure 3.1), the star Sun with its planetary system that includes a planet, call it Earth, in the habitable zone. More thorough research might indicate advanced forms of life, in particular a species with some signs of intelligence. Let us call them Vialacteans. It seems that Vialacteans have just appeared on Earth: there are only a few scattered tribes. They have not had time yet to develop a civilization as advanced as ours.

Which stage of development have Vialacteans reached today? Light and information that reach us now are 2.5 million years old. They appear to be a fragile form of life and they have thus probably disappeared since then. Or there is a slight chance that they have developed into an advanced society, just as we Andromedians have.

Why not send them a greeting message that will take another 2.5 million years? That leaves them 5 million years to vanish … or develop a sustainable civilization and the technology to receive our message. And it will take another 2.5 million years to receive their answer. …

There is probably another way to meet these Vialacteans, but, this time, it requires very long-term sustainability of our Andromedian society. The Via Lactea Galaxy is moving closer to us at a velocity of 110 km/s. Our own astrophysicists predict that the two galaxies will collide in about 4 billion years to merge into a single elliptic galaxy. Simulations have been made of this extraordinary event (Figure III.2).

It is difficult at this time to predict what would happen to our own star and to the Sun. Some think that we would be rather protected in

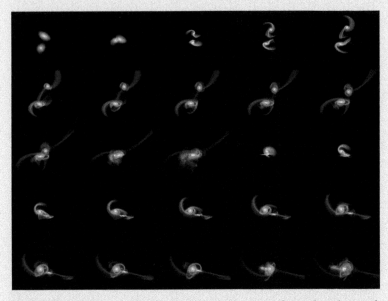

Figure III.2 Simulation of the collision between the Andromeda Galaxy and the Milky Way (the individual pictures are separated by 170 million years).
© John Dubinski / University of Toronto.

the vicinity of our star. The big operations would take place at distances much larger than the size of our planetary system, and their gravitational effect on us would be negligible. It would all translate into magnificent night skies with two large milky strips of stars. When this cosmic ballet comes to an end, will our star and the Sun end up neighbours? Or will one of the two be ejected from the brand new galaxy?

And will our own Andromedian civilization survive during these 4 billion years to be able to observe the phenomenon? The way things go, one may have doubts. …

Let us abandon the Vialacteans to their fate and consider the sky beyond the Via Lactea Galaxy. We observe at larger distance well-structured galaxies, of spiral or elliptic type, then less luminous (because they are further away) and more irregular (because they are younger) galaxies. Figure 3.3 presents a picture taken by the space telescope launched some years ago by NASA, the National Andromedian Space Agency, where one recognizes these different types of galaxies.

I imagine that, if Vialacteans or others have developed similar space capabilities, the pictures they obtain are rather similar. It might seem, but only apparently, in contradiction with another observational fact: each observer, placed at a particular point, sees in a certain way his own version of the Universe since the period at which he sees each astrophysical object depends on this position (today I see the Andromeda of now, the Milky Way of 2.5 million years ago, etc.). However, the huge number of objects in the Universe and the homogeneity of this Universe at large scales imply that each isolated observer can comprehend the Universe in its largest dimensions. Thus, an astrophysicist who is interested in the collision of galaxies can find diverse places in the Universe where she can observe images of pairs of galaxies at various stages of a collision. By putting them one in sequence, she can reconstitute the movie of what is a typical collision/merger of galaxies (Figure 3.4).

Our great Andromedian mathematician and thinker, Blaise Pascal (1623–62), superbly summarized this apparent contradiction in his *Pensées*: 'In space, the Universe comprehends me and embraces me as a point; in thought, I comprehend it.'

4

Inner Space, Outer Space: So Close and So Far Away

Thanks to observation, we have retraced the history of the Universe back to 380,000 years after the Big Bang. With the age of the Universe estimated at 13.8 billion years, it seems that we covered the essential part. Yet the history of what we call the primordial Universe, that is, the first 380,000 years, is very rich. Actually, as we are going to see in this chapter, the essential occurred in a blink, those 'first three minutes' that provided the title of an excellent book by Steven Weinberg.

This rather surprising statement is due to the fact that our measuring time in seconds or in years is not adapted to the description of the primordial Universe: after all, the second is based on our pulse.[1]

In order to understand it better, we will be experimenting with gravity with one of my favourite devices, a roller coaster. Imagine yourself at a fairground and let us board one of these roller coasters. Our car slowly climbs to the top and then off we go … wheeze! … and we are already at the bottom, before climbing up again. You will agree that a lot has happened in the few seconds of almost vertical drop. And I am not talking here about psychological time only.

Let us analyse closely our motion. When our car is first climbing slowly up the ramp to the top, it is transforming electric energy into gravitational energy. In the attraction field of the Earth, the gravitational energy is directly proportional to the difference of height. Once on top, the rest of the ride will just be spending our gravitational energy capital: we will turn it into velocity (kinetic energy) and friction on the rail tracks (that will slightly heat up). In Figure 4.1, I have represented the ride using height (a), car velocity (b), and gravitational energy (c) as a function of time.

---

[1] It is said that Galileo was measuring time with his own pulse, for example in his famous pendulum experiment.

You can see how similar the height curve (a) and the gravitational energy curve (c) are, for the reason just given. After the first drop, we climb again but not as high: some of our gravitational energy was dissipated as friction on the rails. And similarly for the last bump, after which we coast away at slower and more or less constant velocity.

Precisely, if we look at the velocity curve (b), we see that our velocity is constant as we climb up, and suddenly increases as we fall along the first slide: our motion is accelerated and, at the bottom, we have gained kinetic energy out of our gravitational energy capital. But by climbing up the second bump we partly restore this gravitational energy and slow down.

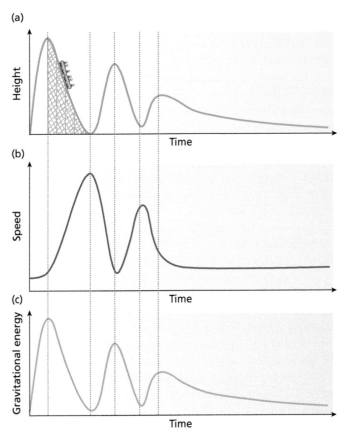

Figure 4.1 The roller coaster ride as seen using the evolution of height (a), car speed (b), and gravitational energy (c) with time.

Well, the evolution of the Universe follows very much the first slide of our roller coaster, and we will see that its energy varies with time much like the gravitational energy of Figure 4.1c in the first part of the ride.

How do we measure the cosmic evolution of the Universe? We have seen that we can use the average temperature, which is decreasing with the expanding Universe. We can convert this temperature into energy units, since heat is a form of energy. This is the traditional approach that I will follow here. Since we are going to make contact with the physics studied in particle accelerators, I will use the energy unit in this field, the electronvolt (eV).

The electronvolt

The electronvolt is the kinetic energy acquired by an electron moving in an electric circuit connected on a battery of 1 volt. The Large Hadron Collider (LHC) at the European Organization for Nuclear Research (CERN) accelerates particles up to energies of 14,000 GeV, the gigaelectronvolt (GeV), corresponding to 1 billion eV. The temperature of the Universe at hydrogen recombination is 3000 K, which corresponds to an energy of 0.26 eV. The present temperature/energy of the Universe is 1100 times smaller, i.e. 0.00024 eV.

The exploration of the approximately 14 billion years that separate the period of recombination from today makes it possible to explore the Universe between 0.00024 and 0.26 eV. In contrast, the exploration of the first 380,000 years, i.e. before recombination, allows us to explore the Universe beyond 0.26 eV. In particular, the energy of the LHC at CERN (14,000 GeV) is reached 10^{-15} seconds after the Big Bang. As you can see from Figure 4.2, the evolution of the Universe is one big slide, and the greatest fun is during the first three minutes. Wheezz!

Since we are comparing the energy scales in accelerators and in the primordial Universe it is timely to discuss now the link between the two approaches. We often read that conditions reproduced at CERN are close to those of the Big Bang. This is not exactly true because, let us remember, the primordial Universe was both hot and dense. A particle collider reproduces, by accelerating particles, the conditions of energy observed in the primordial Universe but certainly not the conditions of density. It remains true that, in these accelerators, new particles are produced at high energy, particles which were also produced by thermal fluctuations

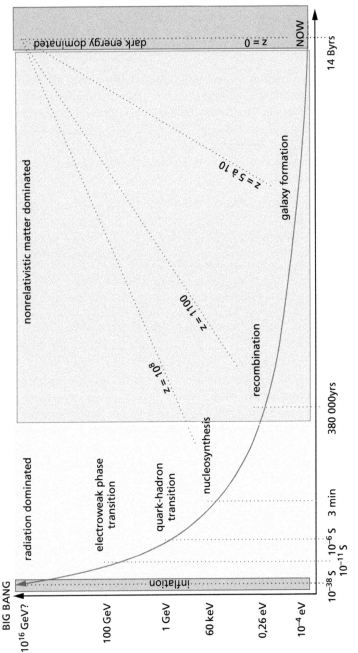

Figure 4.2 The cosmic slide: main stages of the history of the Universe in terms of time passed since the Big Bang (horizontal scale) and energy (vertical scale in eV and its multiples) and in red shift (z, following the diagonal dotted lines). The periods are marked in different colours depending on the energy component then dominating the evolution of the Universe: vacuum energy in red, radiation energy in yellow, and mass energy of particles in green.

in the primordial Universe. It is in this sense that accelerators give us information on these very ancient epochs and that the study of the infinitely small gives us information on the infinitely large Universe. Conversely, the study of the primordial Universe is a remarkable laboratory for microscopic physics. Will the first moments in the Universe be those where we can reconcile the two infinities, inner space and outer space?

Let us resume our voyage into the past from where we left off, at recombination 380,000 years after the Big Bang. I will identify epochs not in reference to the present time (they are all about 14 billion years ago) but in terms of the time elapsed since the Big Bang, or otherwise in terms of explored levels of energy, which is equivalent (Figure 4.2 provides the translation).

At the time of recombination, the most common form of energy in the Universe was mass, whether of ordinary or dark matter. This is not true for the first 40,000 years after the Big Bang. In those very early times, radiation dominated, almost from the time of the Big Bang. This translates as a slightly less rapid evolution of the expansion of the Universe.

But whether the Universe's evolution is dominated by radiation or by mass, one property remains: the expansion decelerates as the Universe evolves. Let us recall our example in the fairground: once the car reaches its maximum velocity, friction slows its velocity down; it then decelerates (on average) all the way to the end. Conversely, when we move backward to the past, expansion becomes more and more rapid.

Element synthesis

An important stage happens between 10 seconds and 20 minutes after the Big Bang that is at energies between 0.01 and 0.0001 GeV (or temperatures between 1 and 0.01 billion K). It is the synthesis of matter, more precisely the synthesis of atom nuclei: in technical terms this is called nucleosynthesis. Again, the name of George Gamow is associated with the discovery of this important phase of the Universe's evolution.

The structure of matter

Ordinary matter is made up of atoms: each atom consists of a central nucleus (of positive electric charge) that focuses most of the atom mass, and of much lighter electrons (of negative charge). The nucleus itself is made of protons (positively charged) and neutrons (of vanishing electric charge). Finally, protons and neutrons are each formed of three quarks. It is the strong force that binds quarks within the proton (or the neutron) and neutrons and protons within the atomic nucleus. As for the electric force, it binds (negative) electrons to the (positive) nucleus.

If we follow cosmic time, we see conversely matter structuring itself from a 'soup' of quarks, gluons, and electrons.

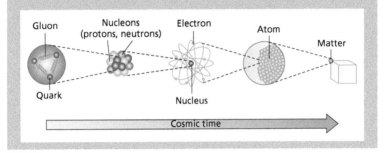

In the primordial Universe, the energies at play were sufficiently high to unbind the structure of atoms and even of nuclei: we get a 'soup' of elementary particles, electrons, protons, neutrons, or even quarks. As the temperature decreases, protons and neutrons form more and more complex ensembles. Thus, a proton bound to a neutron forms a deuterium nucleus, the process of assembling, called nuclear fusion, continuing then towards heavier and heavier nuclei.

Physicists, who know well the nuclear reactions contributing to this process, have shown that, at the end of this phase, 8 per cent of nuclei, that is, 25 per cent of the mass in the Universe, were in the form of helium-4. When the temperature got even lower, the energy available was no longer large enough to allow these fusion reactions: the

A cosmic Lego

From the building blocks that are the proton and the neutron, we may obtain, through nuclear reactions, a deuterium nucleus (1 proton and 1 neutron), then a helium-3 (2 protons and 1 neutron), a helium-4 (2 protons and 2 neutrons), and even a beryllium-7 (4 protons and 3 neutrons) or a lithium-7 nucleus (3 protons and 4 neutrons). The most stable of these nuclei is helium-4.

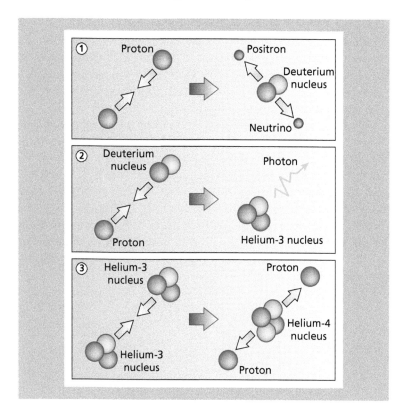

fractions of the different elements were from then on fixed. The predictions for helium-4 were confirmed by observation and constituted one of the first quantitative successes of the Big Bang model. The situation is a little less clear for lithium-7 where different observations give results that are not in agreement. It is fascinating, however, to realize that all the lithium in batteries or of lithium salts used for diverse treatments was synthesized during the first minutes of the evolution of the Universe. It makes it even more precious.

What about heavier nuclei: carbon, oxygen, iron, ...? Physicists believe that they were synthesized later in the Universe's evolution. Indeed, no stable nuclei exist with precisely eight constituents: this constitutes a bottleneck that does not allow us to play with our cosmic Lego beyond seven nucleons. The synthesis of heavier elements is realized in the core of stars where, the density being larger, triple collisions of helium-4 nuclei take place that produce carbon and make it possible to go beyond the bottleneck. Heavy nuclei are thus produced by nuclear reactions within stars and then scattered into the environment when the star at the end of its life explodes into a supernova. Thus, all the iron present on Earth was synthesized within stars and then dispersed in their explosion. All the carbon is as well. It is in this sense that we are made of stardust.

The success of the predictions concerning the synthesis of light elements, in particular of the helium fraction in the Universe, lets us believe that we understand well the history of the Universe from one minute after the Big Bang onwards. In particular, we have a good estimate of the expansion rate at that time: if it had been more rapid (slower), it would have diluted (increased) the fractions computed. For earlier periods, we do not have as precise an observation and our vision of the Universe is more speculative. It mostly relies on the Standard Model, the theory of elementary particles that was built over the past 50 years and is summarized in Figure 4.3.

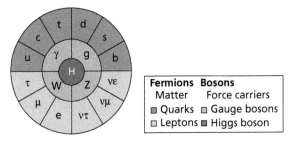

Figure 4.3 The fundamental particles of the Standard Model. Six quarks (u and d, which are the constituents of the proton and neutron, and c, t, s, b) and six leptons (three neutrinos ν_e, ν_μ, ν_τ, electron e, and its two heavier cousins μ, τ); and the four bosons that are the mediators of forces: photon γ (electromagnetic force), gluon g (strong force), and W and Z bosons (weak nuclear force). The Higgs particle H is the only scalar field.

The Standard Model of nongravitational fundamental interactions

We have known since the 1960s that protons and neutrons are made of three quarks. These quarks are *confined* inside the proton or the neutron by the strong force or strong interaction: they do not exist as free particles. The theory of this strong interaction was elaborated in the 1970s: it is quantum chromodynamics (QCD) that forms one of the building blocks of the Standard Model of fundamental interactions.

In this theory, the mediators of the strong force are gluons, analogues of the photons, mediators of electromagnetic interaction. Quarks interact by exchanging gluons. Gluons are responsible for the confinement of quarks, i.e. the fact that quarks do not exist in isolation. And since gluons interact with themselves, they are confined as well.

The other side of the Standard Model concerns the electromagnetic force (mediated by the photon) and the weak nuclear force, responsible for beta radioactivity whose mediators are the particles W and Z, (intermediate force carrier bosons) discovered at CERN in 1983. Actually, the Standard Model unifies these two fundamental forces and shows that, as different as they may seem, they are two sides of the same coin, i.e. the complementary aspects at low energy of a unique force at high energy, the electroweak force. For this purpose it is necessary to introduce the missing link, the Higgs field: at high energy, this field is vanishing and electromagnetic and weak forces have the same characteristics (the electroweak force); at low energy it has a nonvanishing value and electromagnetic and weak forces are different.

These forces are exerted between the known 12 elementary constituents of matter: six quarks, and six particles known as leptons, the best known of which are the electron and the neutrino (Figure 4.3). I might note that the Standard Model carefully avoids dealing with the fourth fundamental force that is at the focus of this book, gravitational force.

The Standard Model received its ultimate confirmation at CERN in 2012 with the discovery of the Higgs particle, associated with the Higgs field (just as the photon is associated with the electromagnetic field).

Fermions and bosons

$$1 + 1 = 2 \qquad 1 + 1 = 1$$

You may recognize in Figure 4.3 the particles that form the building blocks of common matter (quarks u and d that form protons and neutrons, and electrons) and photons (γ), which are the constituents of light. But why is light so different from matter, if they are both made of particles? This is a deep question whose answer relies on a fundamental distinction at the level of particles: quarks and electrons are fermions, whereas the photon is a boson.

To understand the difference, we must be more precise with our statement 'light and matter are obviously different'. In what way? Well, let us take a box of sugar cubes: I can put the sugar cubes side by side; in other words I can juxtapose them. However, I cannot superpose them, that is, move two sugar cubes so that they coincide at exactly the same point. On the other hand, I can easily superpose two light rays: I simply get a more intense light ray. It is this property, related to the behaviour difference with respect to juxtaposition/superposition, that makes a fundamental difference at the microscopic level.

This property was expressed as a principle by Wolfgang Pauli: the elementary particles, which form the building blocks of matter, cannot be in the same microscopic (quantum) state. They are called *fermions*. The other elementary particles, such as the photon, which can be accumulated in the same microscopic (quantum) state, are called *bosons*.

Hence, by accumulating many identical photons we can form a powerful laser beam, or we can generate a strong magnetic field. We can thus generate macroscopic forces. This is why bosons are associated with fundamental forces: their exchange in large numbers between two bodies corresponds to a force (also called interaction) between these two bodies. Just as the exchange of photons generates an electromagnetic interaction, the exchange of gluons between two bodies generates a strong interaction, and the exchange of intermediate bosons W and Z generates the weak force. We say that gluons mediate the strong force, whereas W and Z mediate the weak force.

To summarize, the particles that are the constituents of matter are the fermions. The particles that mediate interactions, or forces, are the bosons. And there is a simple mnemonic to remember their microscopic properties: in the language of fermions, $1 + 1 = 2$, whereas in the language of bosons, $1 + 1 = 1$.

Antiparticles

Matter is well known to us; it is made of quarks and leptons. Antimatter was predicted by Paul Dirac in 1931 when he tried to write the quantum theory of the electron. Since then, it seems to have exerted a fascination on the general public. However, it is as boring as ordinary matter: as shown by Dirac, for each particle there is an antiparticle, with the same characteristics (mass, angular momentum) but opposite electric charge,[2] which

---

[2] This leaves the possibility that neutral particles are their own antiparticles. This is the case for the photon, for example. It is still an open question whether neutrinos are their own antiparticles.

means that the diagram of Figure 4.3 should be mirrored by a similar diagram of antiparticles. The antiparticle of the electron, called the positron (because it is analogous to the electron, but with a positive electric charge), was the first antiparticle discovered by Carl Anderson in 1932, which confirmed Dirac's prediction. If a particle encounters its own antiparticle, they annihilate in a burst of energy (often in the form of photons), which leaves them little chance for coexistence. When antiprotons are produced at CERN to feed them into the accelerator, extraordinary precautions must be taken to prevent them from crossing the path of a proton!

Antiparticles play an important role in what are called *quantum fluctuations*. Quantum mechanics establishes a deep connection between energy and time. In particular, violations of the law of conservation of energy are allowed if they last a very short time: the more violation there is, the shorter it must be. Imagine now that we are in a vacuum, and that a pair formed by a particle and its antiparticle appears (Figure 4.4): their total charge is zero, but they have the same mass m and their total mass energy is $2mc^2$. From a classical point of view, such a process, i.e. the spontaneous appearance of an energy $2mc^2$, is forbidden by the conservation of energy. From a quantum mechanical point of view, this is allowed for a microscopic moment. The larger this energy violation, the shorter this moment. What happens then? The pair annihilates back into the vacuum! The particle and the antiparticle are said to be *virtual*: they have hardly appeared before they vanish. The complete

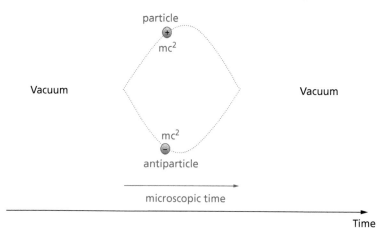

Figure 4.4 When energy conspires with time: a quantum fluctuation formed by a pair of virtual particle–antiparticle.

event is called a quantum fluctuation. A vacuum is the source of perpetual quantum fluctuations. You would never have suspected that a vacuum could be so eventful!

The enigma of the absence of significant antimatter in the Universe

The Universe that we observe is made of matter (with traces of antimatter produced in energetic phenomena). This is good news for us; otherwise, we would be surrounded by explosions of matter annihilating with antimatter. But if microscopic physics treats particles and antiparticles on an equal footing, how is it that we do not see equal amounts of matter and antimatter in the Universe? Was it always this way? Probably not, because the early Universe was a soup of elementary particles subject to the laws of microscopic physics. The general consensus is that the unbalance observed between matter and antimatter is the result of the Universe's evolution. What is fascinating is thus not antimatter itself, formed of antiparticles, but rather its disappearance in the course of the Universe's evolution.

It was first imagined that a transition happened at some early time with one part of the Universe occupied by matter and another by antimatter (just as oil and vinegar separate in a bottle when we stop shaking them), but the frontier between the two regions would be the place of matter–antimatter annihilations, which are not observed in the Universe. It is more probable that a slight imbalance between matter and antimatter happened at some time, which the Universe's evolution quickly amplified until there was the quasi disappearance of any trace of antimatter. Soviet physicist and Nobel Peace Prizewinner Andrei Sakharov stated the conditions necessary for such a differentiation to happen. We don't know, however, at which moment of the Universe's evolution this could have taken place. But this disappearance allowed structures of matter to develop henceforth without the risk of annihilating against structures of antimatter.

Higgs field or Higgs particle?

Let us now meet the prince of elementary particles: the Higgs. But is it a particle or a field?

First, what is a field? In everyday language a field is a planted surface; in other words, at every point in a field we can provide information on what can be found there: a seed, a weed, soil, or stone. For the physicist, a field is an ensemble of data of a physical quantity in a given region of space: for example, a field of pressure or velocity in a water pool, or an electric field on a surface or in a volume, or a gravitational field in the vicinity of a star.

A wave is nothing other than a field in motion; thus, an electromagnetic wave is characterized by the data of an electromagnetic field in every point of a region of space *and for successive times*. It is, thus, a field in *space–time*. But, as we saw at the end of Chapter 3, in quantum physics every wave associated with a fundamental force can be interpreted in terms of particles: the electromagnetic wave, in particular light, can be seen as a superposition of grains of light or photons. There is a duality between the fields of space–time and particles. This is why the domain of relativistic quantum physics, which deals with particles in a relativistic context, is also called *quantum field theory*.

There is, thus, a duality between the photon particle and the electromagnetic field. There is similarly a duality between the Higgs particle and a field, which is naturally called the Higgs field. The term 'duality' is used here to signify that, depending on the physical situations, the unique entity Higgs manifests itself in the form of a field or in the form of a particle, just as the unique entity light can appear in the form of an electromagnetic wave/field or a photon.

The discovery of the Higgs particle at CERN in 2012 thus coincided with the discovery of the Higgs field. But this field has in the context of the standard model specific properties that distinguish it from all other known quantum fields: it is a scalar field. To understand this notion let us return to the electromagnetic field: it is given at each point not only by the value of the electric and magnetic fields but also by their directions (for example, the direction of a magnetic field is given by the compass needle); this is what we physicists call a vector, and the electromagnetic field is called a vector field. In contrast, the Higgs field is characterized only by its magnitude; there is no associated direction. It is not a vector field; it is called a *scalar field*. To understand the difference, let me use a meteorological analogy. Figure 4.5 gives two different kinds of meteorological data over Europe. On the left, you find the velocity and direction of the wind, represented respectively by the size of the arrows and their *direction*: this corresponds to a vector field. On the right, you find the data corresponding to a single quantity, the pressure, through lines of constant pressure: this is a scalar field.

This has an unexpected consequence: the Higgs field may have a constant value throughout space–time. The electromagnetic field could not because of its vector nature: this would imply a privileged direction in space that is not observed in nature.

Gravity!

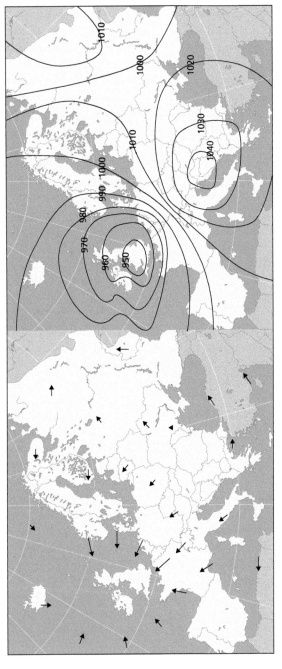

Figure 4.5 Maps of vector (on left, wind velocity) vs scalar (on right, pressure).

The Standard Model uses this property: the Higgs field has a constant value throughout space–time. Does this mean that there are Higgs particles everywhere? Not necessarily because a particle is, in fact, nothing else than a fluctuation localized in space and time of the corresponding field. These fluctuations can be of a quantum nature, in which case they have microscopic lifetimes (the virtual pairs that we described earlier) or of a purely classical nature: they then correspond to the particles observed in detectors.

Quantum vacuum and Higgs field

Let me focus on the notion of quantum vacuum. We will return to it several times in the next chapters. In everyday language, a vacuum is what remains when everything has been taken away. Not much thus, only a space–time framework in which we can later reintroduce particles, atoms, molecules, etc. In a quantum world, can we take everything away? Not really because there are always virtual particles appearing and disappearing: I cannot separate a vacuum from these microscopic fluctuations. This is why it is preferable to identify the quantum vacuum as the fundamental state, that is, the state of minimum energy. A nonvanishing energy for the vacuum? Again, this seems far-fetched from a classical point of view: nothing has no energy! But the quantum vacuum is irremediably associated with quantum fluctuations. These come and go but at any given time, some of them are present and, on average, their individual microscopic energies add up to a nonzero value: this is the energy of the quantum vacuum.

To allow a better understanding of this notion, I will take an example from the world of movies. On a film set, after having recorded the different scenes corresponding to the specific set of the day, the sound engineer asks the whole team to stay quiet for a couple of minutes in order to record the 'silence'. This silence will allow later to edit the rushes: it is the background that will allow the engineer to reconstruct if necessary the sounds over it. It is characteristic of the set, of the volume of the room, of the people present (actors and technical team), of the weather on that day. It is an absence of sounds but it is full of sound fluctuations.

To return to the Higgs field, the central role that it plays comes from its coupling to the other fields/particles. All the other particles of the Standard Model have a mass because they interact with the Higgs field and because this field has a nonvanishing value throughout the

Universe. We started initially to track gravitation to understand the concept of mass, and it is microscopic physics that opens up unexpected horizons. Since the gravitational force is outside the framework of the Standard Model, it is of course the *inertial* mass that we are talking about. The inertial mass of a particle is, in the Standard Model, proportional to the value of the Higgs field in the quantum vacuum (the same value for all particles) and to the intensity of its interaction with this Higgs field (what we call its coupling to the Higgs, characteristic of the particle). Why is the electron 6 million times lighter than the top quark? Its coupling to the Higgs is 6 million times smaller. It is also the coupling of the intermediate bosons W and Z, the mediators of the weak force, to the Higgs that ensures they have a nonvanishing mass.[3]

Phase transitions

Around 10^{-6} seconds after the Big Bang, the Universe went from a phase of quarks and gluons, all gathered close together, to a phase of well-identified particles, i.e. a phase of quarks and gluons bound together into protons and neutrons. Calling this process a change or transition of phase is well suited: we name, for example, the transformation from liquid water into vapour or from ice into liquid water a phase transition.

Another transition occurred at a much earlier era. It is directly connected with the Higgs field. The *electroweak phase transition* happened when the temperature of the Universe was on the order of 10,000 billion K (10^{13} K if I use powers of 10, corresponding to an energy of around 100 GeV), which happened 10^{-10} seconds after the Big Bang. Beyond this temperature, i.e. at earlier epochs, the electromagnetic force and the weak nuclear force become a unique force, the electroweak force (Figure 4.6). This is related to a singular behaviour of the Higgs field at the phase transition: its value in the vacuum becomes zero above the transition temperature. It follows that all particles become massless. In particular, the masses of the intermediate bosons W and Z, mediators of the weak force, vanish, and thus W and Z appear very similar to the massless photon. The weak force then becomes very similar to the electromagnetic force: we talk of the unification of forces and of symmetry restoration; differences between particles become blurred. Thus, as we

---

[3] This nonvanishing mass is responsible for the microscopic range of the weak nuclear force.

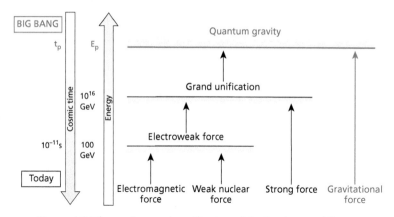

Figure 4.6 The road towards unification of the fundamental forces.

get closer to the Big Bang, not only the structures crumble but the differences of behaviour become less marked and one goes towards more symmetry.

We suppose, but for the moment without any experimental confirmation, that this march toward unification continues, and that at an energy on the order of 10^{16} GeV, strong force and electroweak force turn into the incarnations of the same unified force. This is what physicists call the grand unification phase transition (Figure 4.6).

Towards the Planck energy and quantum gravity

We have until now scarcely mentioned gravitation, and we have given the central role to the other fundamental forces described by the Standard Model. This has allowed us to identify some key periods in the evolution of the Universe, even though gravity remains the engine of this evolution, as we have seen in the preceding chapters.

But, as we get back to the Big Bang, should the road towards unification include gravitation? And if the answer is yes, from which time? To answer this question, it is necessary to have a *quantum* theory of gravitation that would allow us to connect with the *quantum* theory of the Standard Model. There are for the moment theoretical efforts in this direction but no specific prediction that could be experimentally verified. We will come back to this central question. Let us content ourselves here with outlining the problem.

First of all, it is quite possible that there is no quantum theory of gravitation, i.e. that gravitation is by essence a nonquantum theory. If this is the case, it is, however, difficult, given the close connections between the structure of space–time and gravitation, to understand in which way the dynamics of space and time could be modified when we get close to the Big Bang and to explain the theoretical difficulties that we are facing.

> ## Galileo and dimensional analysis
>
> Let us illustrate Galileo's method on a famous example that we already encountered at the beginning of this book: the observation of the period of oscillation of the chandeliers of the Pisa cathedral, which put him on the track of the universality of free fall. Take a chandelier of mass m, measured in kilograms, attached to a rope of length l measured in metres. The acceleration due to Earth's gravitation is $g = 10$ m/s^2. How do we express the period of oscillation, that is, the time, measured in seconds, between two oscillations in terms of these quantities? A little thinking will convince you that l/g is measured in seconds squared; hence, the period, measured in seconds, is proportional to the square root of l/g, and does not depend on the mass. This is exactly what Galileo observed in the cathedral, and later in his experiments.

It is Galileo who taught us the power of dimensional analysis when we are facing a new problem and we do not have the full theory. We should identify the relevant parameters and then combine them in such a way that they have the dimensions (i.e. the units) of the searched quantity.

Following Galileo's example, we can try to identify the energy scale of quantum gravity (Figure 4.7). For this purpose, we can make use of Newton's constant G, which characterizes gravitation. Planck's constant h characterizes quantum physics. Finally, the velocity c of light is characteristic of relativity. We can easily construct a combination of these three constants that is measured in metres: it is called the Planck length and its value is 10^{-35} m. We can also write another combination of the three fundamental constants measured in seconds: this is the Planck time or 10^{-43} s, which gives an order of magnitude of the duration of the Planck era, just after the Big Bang when the Universe's evolution was governed by quantum gravity. Finally, one last combination has the dimension of an energy; i.e. it is measured in m^{2}kg/s^2: it is called the Planck energy and has a value of 10^{19} GeV.

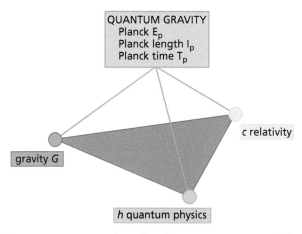

Figure 4.7 How to construct the scales of quantum gravity out of the fundamental constants associated with gravitation (G, measured in $m^3/(kg.S^2)$), quantum physics (h, measured in $m^2.kg/s$), and relativity (c, measured in m/s).

The theme of the connection between gravity and quantum physics will be recurring throughout the rest of this book and the discussion will become richer as we go along. For the moment, I will only introduce a newcomer: the graviton. Indeed, just as the electric force between two charges can be interpreted in a quantum context as the exchange of photons between the two charges, the gravitational force between two masses can be seen *from a quantum perspective* as the exchange of particles called gravitons. To the photons, grains of light, correspond the gravitons, grains of gravitation.

And just as light is both a collection of photons and a wave, we can say that gravitational fields can be assimilated into a collection of gravitons, or into waves, the famous gravitational waves.

The Big Bang

El mundo era tan reciente, que muchas cosas carecían de nombre, y para mencionarlas había que señalarlas con el dedo.

(The world was so recent that any things lacked names and, in order to indicate them, it was necessary to point.)

GABRIEL GARCIA MÁRQUEZ, *One Hundred years of Solitude* (1967)

Everything started with a joke.

Cosmologists developed in the years 1930–40 a model of a Universe that became hotter and denser as one moved back in time. At a certain point, the equations of the physicists indicated that temperature and density became infinite. Was this a sign that their equations were incorrect or that one was reaching the limit of time? The cautious physicist would choose the first solution, but it is tempting to lean towards the second that partakes of the myth of the origin, dear to mankind. The Universe, and space–time itself, would have appeared in a primeval explosion and then evolved according to the principle of an expanding Universe, gradually cooling down. As we have seen, the British cosmologist Fred Hoyle, who favoured a stationary Universe, tried to disparage this primordial explosion by jokingly qualifying it as the Big Bang.

By doing this, he granted this initial moment a large publicity. The image is attractive and the public soon adopted it. Beyond its natural link with the myth of origins, it made it possible to establish a sort of well-defined and reassuring frontier between the scientists' domain (the post Big Bang) and the domain of the unspeakable (the pre Big Bang). From my own experience, I have realized that often a nonscientific public (and even a scientific one) is ready to question the whole acquired knowledge of modern cosmology, except for the Big Bang, of which they have, in a certain way, taken ownership. 'Interfere with everything besides the Big Bang'. Yet, the Big Bang is a physicist's invention; it is only a picturesque term to designate a period of time in the Universe about which we do not have answers to all the questions, and we even have difficulties making these questions explicit. The public is so little used to hearing scientists recognize their ignorance that everyone immediately deduces that the answer lies outside the scientific domain.

Moreover, the topic is touching very fundamental considerations, close to philosophical questions and religious faith.

Let us try to be more explicit about the questions that have been raised.

I will start by immediately discarding a preconceived idea that often prevents one from asking the right questions: no, the Universe was not reduced to a small volume ('a pinhead') at the time of the Big Bang! Are you surprised? Indeed, we will see later that the Universe is most probably infinite. Then, as a young child asked me once, how could the Universe have evolved from the size of a pinhead to an infinite size, in a finite amount of time? Out of the mouths of babes come words of truth and wisdom: the Universe was most probably infinite at the time of the Big Bang.

How to understand this? Is not the Universe contracting when one moves back in time? Indeed, but not to the point of becoming finite! To understand the issue, imagine that you have at your disposal an *infinite* rope (thus with no end) stamped with equidistant white marks. Now suppose the length of the rope is contracting with time: every second the distance between two adjacent marks is divided by 2. Thus, after 3 s, the distance between these two marks is divided by $2^3 = 8$, after 10 s by to $2^{10} = 10^{24}$. If you take two marks that were 1 km apart, after 20 s they will be 1 km$/2^{20}$, i.e. 1 mm away from each another, roughly the size of a pinhead. Obviously, the rope is still infinite. In other words, contraction implies that any two points get inexorably closer, but not that infinite is turned into finite.

In fact, the confusion originates from diagrams such as in Figure II.3 where the Big Bang appears as a luminous point. Such diagrams should be understood not as representing the whole Universe but only the part accessible to our observations today. Indeed, this part is finite: since the Universe is 14 billion year old, light could only have travelled a distance of 14 billion light years since the Big Bang. Just as the kilometre of rope in the above example, this sphere of 14 billion light years was contracted into a very small region immediately after the Big Bang. One thus recovers the traditional picture. But you see that our vision of the global Universe is changed: for example, in these early times, we were very close to another pinhead-sized region, that developed into another sphere of 14 million light years, with which we have no contact.

We saw earlier that the singular behaviour of the Big Bang (infinite temperature and density)—physicists call this a singularity—appears when one moves back in time. Should we go back this far or have we trespassed the limit of validity of our theory? In fact, infinite temperature means infinite energy, hence an energy larger than the Planck energy: we have reached the quantum gravity regime of the Planck era for which we do not have currently a satisfactory theory. In other words, the theory at our disposal is no longer valid at energies larger than the Planck energy. In time scale, this corresponds to a time smaller than the Planck time (10^{-43} s). Is the Big Bang singularity only due to the fact that we have not found the right theory at such scales of time and energy? Or is it inherent to the problem of origins? Only more advanced studies will tell us.

The effort concerns of course the construction of theories unifying all known fundamental interactions, including gravity. A theory that has raised many expectations is string theory, a quantum theory that identifies the most elementary entities not as point particles but as one-dimensional objects called strings. The size of such a microscopic string introduces a new fundamental distance scale, from which we can build a time scale or an energy scale. Particles can then be described as oscillation modes of these strings, or, if you prefer, harmonics of these microscopic strings. When the theory was formed in the 1970s, Joël Scherk and John Schwarz realized that one of these particles had all the right properties to be a graviton: it was thus a potential quantum theory of gravitation. One of the most surprising aspects of string theory is that internal coherence at the quantum level imposes a number of space dimensions larger than 4. This means in particular that when one gets near the energy scale of strings (very close to the Big Bang), one identifies new space dimensions! In what follows, we will return to string theory, its successes and its shortcomings encountered, despite more than 40 years of hard work performed by a large community of theorists.

Without focusing on a particular theory, let us recall the close connection between the theory of gravitation and the structure of space–time. In order to conceive a quantum theory of gravitation, it is very probable that we must deeply modify our understanding of space and time. To this effect, recall the beginnings of quantum theory in the first years of the twentieth century. For hundreds of years, a large majority of scientists had considered matter as a continuous medium: we just have to look at objects around us to convince oneselves! However, here was the atomic structure of matter and the fact, which clashes with our common perception, that material objects are essentially made of vacuum, since matter is concentrated into atomic nuclei of microscopic dimensions (even compared with the size of the atoms themselves).

What will happen with the notions of time and space in a future theory that will unify gravitation and quantum physics? Will we have to associate with them a discontinuous structure of grains of space and time as we had to do for matter? Maybe not, but it is probable that, very close to the Big Bang, we will have to replace these notions that are so familiar to us with others just as unfamiliar as atoms were to the contemporaries of the early twentieth century. We only have vague ideas or embryos of theories on these notions, but they have remarkable consequences on our global vision. Indeed, they mean that space and time as we experience them currently are notions that emerged, probably at an era close to Planck time. Is there in these theories a pre-Big Bang? It is probable that the question cannot be asked in such terms since the notion of time itself (and thus of 'before' and 'after') is modified in these theories.

5

The First Moments:
From Inflation to the First Light

After the flashback of the past two chapters let us move forward in time. We are going to be interested in the period immediately after the Big Bang, a phase called cosmic inflation, during which the expansion was very rapid, exponential to be precise. This quasi-explosive phase is often confused with the Big Bang itself. But the Big Bang is not necessarily connected with an explosion. In fact, it is only a generic term to designate a phase of the Universe's evolution that lies outside the domain of validity of our equations. The phase of inflation is, on the other hand, very well described by laws that we know: it takes place after—probably just after—the Planck era. There is no need for a quantum theory of gravity, even if we will use in what follows notions of quantum physics as well as gravity.

Before we start, I will warn the reader that this chapter and the next one may be a little more difficult reading than the previous ones. They do not need any technical background but the concepts are somewhat more involved and thus more difficult to grasp. Since the remaining of the book depends very little on these concepts, you may go directly to Chapter 7 and sequential chapters, if you are mainly interested in black holes and gravitational waves. You can then return to Chapters 5 and 6 to get a full account of the role of gravity in the Universe.

The scenario of inflation was proposed at the very beginning of the 1980s, under different forms, by Alexei Starobinsky, Alan Guth, Andrei Linde, Andy Albrecht, and Paul Steinhardt to solve a certain number of fundamental problems that the standard model of cosmology (described in the previous two chapters) had to face. These problems had been highlighted by the Soviet physicist Yakov Zel'dovich in the early 1970s. They are mainly the horizon and the flatness problems.

Question of horizon

The notion of horizon is central in cosmology and more generally in general relativity. It is intimately related to the principle of causality, according to which the cause precedes the effect. More precisely, the only events that can influence my present are those found in my past. What is an event? The data of a point in space–time, that is of a date and of a location. For example, the death of the playwright Molière took place *on the 17th of February 1673 at ten in the night at his home in Paris, 14 Rue de Richelieu*: this is an event. *This morning, at home,* I let the milk boil over the pot: this is another event. Is there a cause and effect relationship between these two events? Possibly: for example, I was reading this morning about the death of Molière, and I did not pay attention to the pot of milk. Conversely, the principle of causality tells us that the death of Molière cannot depend on the event 'milk boils over'. I feel relieved …

All this is well known but has consequences at the scale of the Universe that are less known, in particular because there exists a maximal velocity of propagation, the velocity of light. According to the principle of causality, only the events in my past can have an impact on the events that happen to me today. What is my past? It is a set of events about which I can, in principle, collect information today. But, according to the theory of Einstein, the velocity at which information travels is at most equal to the velocity of light. Since the age of the Universe is 14 billion years, this implies that points in the Universe further than 14 billion light years from me are not in my past because information has not had time to get to me (Figure 5.1). We have seen that the Andromeda Galaxy lies 2.5 million light years from us: is what happened this morning in this galaxy, or what happened there on the day of Molière's death, in my past? No, because the information takes 2 and a half million years to reach me.

We note that only a finite region of space–time (which is an ensemble of events in the Universe) is in our past. This region has a frontier that we call, using an analogy, a horizon. What is inside the horizon is in our past and can have an impact on what happens to us today; what is outside cannot. Note that this horizon is a space–time notion, whereas the familiar terrestrial horizon seen by the sailor defines only a region of space.

But, just as for the terrestrial horizon, this region of space–time, and consequently the horizon that limits it, depends on the observer: it is

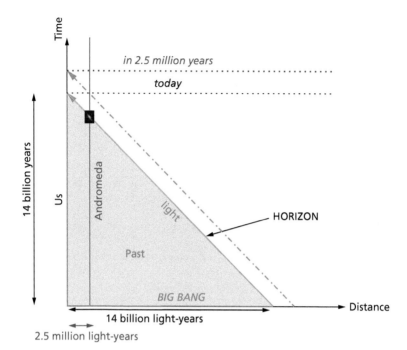

Figure 5.1 Space–time diagram showing the propagation of a light ray emitted at the Big Bang and reaching us today (full line denotes Horizon) or in 2.5 million years (dotted line).

not the same for me and for an observer in the Andromeda Galaxy. It also depends on time: in 2 and a half million years, what happened on the day of Molière's death in the Andromeda Galaxy will enter into my past and can influence the events that will have happened to me.

We now must introduce a further complexity: space–time is expanding. Imagine for a moment that we live on an inflatable Earth: when the Earth was the size of an orange, our terrestrial horizon was a fraction of a centimetre. Nowadays, it is a few kilometres. We conclude that the size of the horizon increases in an expanding Universe.

Conversely, the horizon radius decreases as we go backwards in time. At the time of recombination (epoch corresponding to a redshift of 1100), it was reduced, because the Universe has been in expansion since then: when we look at the sky today, two zones which make an angle of vision larger than $1°$ (Figure 5.2) belonged to different horizons at the time of recombination.

This poses a problem of principle. Indeed, we have seen that the cosmological microwave background (CMB) detected by Penzias and Wilson is very isotropic; i.e. it has exactly the same properties in all directions. But at the time it was produced, the horizon was much reduced. How is it that zones in the sky distant by more than 1°, which

Size of the horizon and Hubble parameter

It follows from the definition that the size of the cosmological horizon grows (linearly) with time: for example, a larger fraction of space–time is in our past (we live 14 billion years after the Big Bang) than in the past of a cosmic background photon at the time of recombination (380,000 years after the Big Bang).

This is, of course, related to the expansion rate of the Universe, measured at a given time by the Hubble parameter. We have seen that the expansion is decelerating: the Hubble parameter decreases with time; more precisely, its inverse is proportional to time. *The size of the cosmological horizon is, in fact, proportional to the inverse of the Hubble parameter*, throughout the evolution of the Universe. And you might guess what is the factor of proportionality: the velocity of light *c*! This should not surprise you: the notion of horizon is intimately connected with the finiteness of this velocity. And Galileo (dimensional analysis) might be called to the rescue: the size of the horizon is distance measured in metres, the velocity of light is measured in metres/second, and the inverse Hubble parameter is measured in seconds.

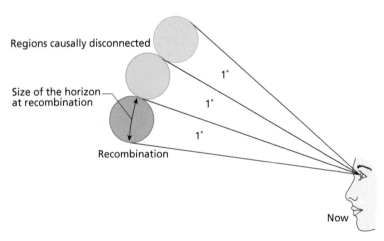

Figure 5.2 The horizon, as seen in the sky, of the photons of the cosmological background at the time of recombination.

at the time of recombination could not have exchanged information since the Big Bang, turn out to have exactly the same properties from the point of view of the cosmological background?

Is the Universe open, closed, or flat?

We insisted in Chapter 4 on the fact that gravity curves space–time. But this was a local effect due to the presence of a large mass. Is space–time as a whole curved? Indeed, the solutions of Einstein's equations that describe our four-dimensional *space–time* (three space dimensions and one time dimension) all correspond to a nonzero curvature. But this does not mean that the three-dimensional *space* is itself curved. We are even going to see in the next chapter that most probably space is flat.

But let me first explain what is a flat space and more importantly what is a nonflat space. First, a warning: in everyday language, we tend to use 'flat' in the sense of a planar or horizontal surface. A surface has 2 dimensions, or coordinates, which make it possible to identify a point on the surface. But space around us has 3 dimensions. Well, physicists and mathematicians use the word in a more general sense. A space is called flat by physicists if the standard laws of Euclidean geometry (the ones that we learn at school) apply. For example, two parallel lines never meet. We have seen that a mass locally curves light rays. So it might not be surprising that, in a nonflat, i.e. curved space, parallel lines are curved and eventually meet. For reasons of commodity, I will explain the different notions with 2-dimensional surfaces for which visualization is simpler but remember that 'flat' does *not* mean a planar 2-dimensional surface.

A two-dimensional flat space is shown in Figure 5.3a. If we draw a triangle on it, old school memories tell us (Euclidean geometry again!) that the sum of the angles of a triangle is 180°. This is no longer true on a curved surface: if we draw a triangle on a sphere, the sum of the angles is this time larger than 180° (Figure 5.3b). This is characteristic of a *closed* curved surface such as the sphere. Let us note that the larger a sphere's radius (curvature) is, the flatter the surface of the sphere is, and the closer the sum of the triangle's angles is to 180°. Are there surfaces for which the sum is smaller than 180°? The answer is yes: those are the *open* curved surfaces, the most classical example of which is the saddle (Figure 5.3c). You might also verify that, in the case of a sphere, two lines that start parallel will eventually converge towards one another, whereas they diverge in the case of the saddle.

(a)   (b)

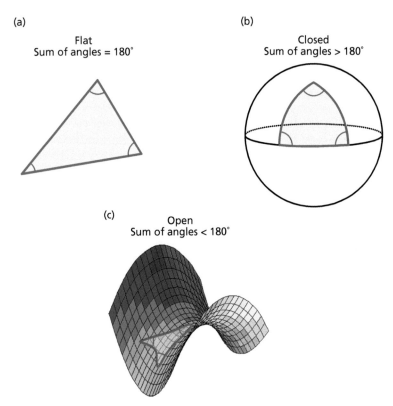

Flat   Closed
Sum of angles = 180°   Sum of angles > 180°

(c)   Open
                Sum of angles < 180°

Figure 5.3 A triangle in a space that is: (a) flat, (b) curved and closed (sphere), (c) curved and open (saddle).

To verify that you have understood, here is a question: is the surface of a cylinder flat or curved, open or closed? To answer, it is enough to draw a triangle: take a sheet of paper, draw a triangle on it then glue together the opposite sides to turn it into a cylinder; you obtain a triangle and the sum of the angles of this triangle is 180°. You have just discovered that the cylinder has the geometry of a flat surface! It is flat, in the sense of physicists and mathematicians.

Is our Universe spatially flat? In other words, is the associated space flat? If not, is it open or closed? This can have important consequences on the cosmological scenario, because, for a closed Universe, the phase of expansion is followed by a phase of contraction where the Universe tends to a new singularity, a Big Crunch, before starting again with a new Big Bang, and so on in cycles.

To answer this question of flatness, we can imagine that it is enough to draw a triangle in space. If the curvature radius is of a cosmic size, it is very unlikely that we could measure with enough precision the sum of the angles. Or even draw a triangle of cosmic size ...

But there is a more straightforward test. Remember that Einstein's equations connect the geometry with the energetic content of the Universe. It turns out that, for a homogeneous and isotropic Universe such as our own, there exists a very simple relation between the average energy density in the Universe and the spatial curvature: space is flat if the energy density is equal to a value called the critical density; space is closed if it is larger, and open if it is smaller. To be more precise, the value of the critical density today is 10^{-26} kg/m^3. This may not seem big to you. It is indeed only 5 hydrogen atoms per cubic metre, but remember that it is the energy density averaged over the whole Universe; for example, the average density in our galaxy disc is six orders of magnitude (10^6) larger.

In the years 1970–80, all the (luminous and dark) matter identified in the Universe contributed to at most 30 per cent of the critical density. Hence, unless a component of the Universe had been missed, this led everyone to conclude that space is open.

The flatness problem

The critical density has varied during the evolution of the Universe, just as the average density has. In fact, the difference between the two has varied as the inverse of the expansion velocity. Since we have seen that expansion has decelerated during the Universe's evolution, the velocity of expansion has decreased. Hence the difference between the two values should increase in the future and decrease when we go back into the past. This is precisely where the problem lies, because the Universe is ancient and we can take its history back to primordial times. A precise calculation shows that if the average energy density of the present Universe is 30%, i.e. $0.3 = 1 - 0.7$ of the critical density, it was $1 - 10^{-58}$ at the time of grand unification (when the energies at play were on the order of 10^{16} GeV). In other words, space must have been extraordinarily flat at that time. What could be the reason for this?

The inflation era

The problems that we have just discussed, horizon and flatness, seem to favour a scenario where, at least for a period of time, the expansion of the Universe accelerated instead of decelerated. Such a scenario, called cosmic inflation, was proposed by Alan Guth at the beginning of the 1980s.

The idea is rather generic and depends little on the explicit model chosen (grand unification in Guth model). It yields a central role to the energy of the quantum vacuum.

Remember that what we call vacuum is the state of minimal energy. It is constantly filled with quantum fluctuations in the form of particle–antiparticle pairs, which appear then disappear. At any given time, the pairs present confer a nonzero energy to the vacuum: this is the vacuum energy.

In a phase transition, the quantum vacuum reorganizes itself. The fluctuations associated with each of the two phases being of a different nature, the energy of the vacuum is modified. It is necessarily inferior in the second phase (otherwise, the transition would not proceed). If the transition is slow, it is possible that there is a latency and that the Universe remains for some amount of time in the initial vacuum state of higher energy. The vacuum energy thus stored contributes to the Universe's expansion: this behaviour is precisely the one that had been identified by de Sitter as early as 1917, i.e. an exponential expansion of the Universe.

Exponential expansion

Imagine a Universe in such a rapid expansion that, every second (or every unit of time, whatever you choose as a unit of time), distances double. After three units of time, the galaxy that was 100 million light-years from us is 800 million ($2 \times 2 \times 2 = 8$). After ten units of time, it is approximately 1000 times further away (2 multiplied 10 times is 1,024), and so on. This is exactly what is known as an exponential expansion.

Such behaviour was obtained by de Sitter when he solved Einstein's equations with a cosmological constant. As a matter of fact, vacuum energy appears as a constant in Einstein's equations: this is why one recovers the regime identified by de Sitter.

The rate of expansion is constant;[1] this is what we called the Hubble parameter. You may remember that this parameter fixes the size of the horizon. We thus conclude that, during a phase of exponential expansion, the size of the horizon remains constant. This, at first, seems in contradiction with such a rapid growth of distances. But, precisely because of this almost explosive growth, points distant from us recede at velocities larger than the speed of light, the astrophysical objects placed there cannot send us information, and they thus lie outside our horizon. Only points within a fixed distance (the constant horizon) are close enough to be within our causal reach.

The physical distances increase under expansion as exponential functions of time. The expansion *rate* is constant; thus, the expansion accelerates with time.

How does this all end? The transition must be completed, which imposes that there is a built-in instability in the system that takes over after a certain time. The quantum vacuum then catches up with the

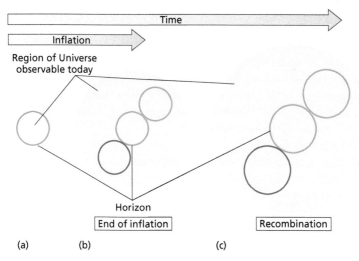

Figure 5.4 Evolution of the size of the region of Universe observed today (in blue) and of the horizon (represented by circles) (a)→(b): during inflation, the size of observable Universe grows rapidly but the size of the horizon remains constant; (b)→(c): in subsequent evolution, they both grow slowly.

[1] But the velocity of expansion increases exponentially with time (the rate is the *relative* variation of distance with time).

state of lower energy (which is the vacuum state corresponding to the second phase). The phase of exponential expansion naturally ends.

It is necessary that this expansion has enabled the distances to enlargen by a factor at least equal to 10^{26} in order for the observable Universe to come out of a unique causal region of the primordial Universe. Figure 5.4 illustrates how the observable Universe and the horizon evolve during and after the exponential expansion: the whole observable Universe was within in a single horizon at the beginning of inflation; all points are thus causally connected today.

Inflation solves the flatness problem

When the expansion is exponential, the expansion velocity increases exponentially with time. Thus, independently of its initial value, the density of energy in the Universe drew, during the phase of inflation, exponentially closer to the critical value of the energy density. One of the predictions is, thus, that this density today is 10^{-26} kg/m^3, and that the Universe is spatially flat. *The inflation phase has, in some way, flattened any spatial curvature.*

A potential difficulty remains: where has all the matter gone? Indeed, the phase of inflationary expansion is so efficient that it has diluted away in a spectacular manner the primordial soup of particles: there remains only a few particles within each causal horizon, not enough to build the 200 billion galaxies that we estimate to be in our observable Universe, that is, within our present horizon.

It turns out that the scenario of inflation ends with a phase called reheating, during which matter is regenerated: it is this new matter that forms our present Universe.

The energy provided for reheating is the difference of energy between the vacuum energies of the two phases (Figure 5.5).

The standard inflation scenario involves a scalar field of the Higgs field type. Such a scalar field, called inflaton, can change its vacuum value throughout the history of the Universe,[2] which corresponds to a variation in the energy of the quantum vacuum. The matter produced

---

[2] Similarly, we have seen that the Higgs field has, in the primordial Universe, a vanishing value, before it acquires, at the electroweak phase transition, the nonzero value that it has today.

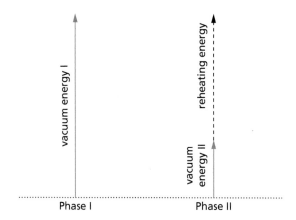

Figure 5.5 Energy budget in the two phases of the inflation scenario.

during reheating and that will form the matter content of the Universe at the end of inflation is produced by the decay of the inflaton field. From then on, the scenario coincides with the standard model of cosmology described in the preceding chapters.

Thus, the inflation scenario helps to solve a certain number of cosmological problems related with cosmology, without endangering any of its successes. But can we test some of its predictions? One of them has been mentioned: the scenario predicts a Universe today that is spatially flat; i.e. the average energy density is 10^{-26} kg/m^3. Observations until the end of the 1990s were in disagreement with this prediction. But, as we will see in what follows, it has been since discovered that an energetic component of the Universe had been ignored.

Another prediction is the presence of anisotropies in the cosmic microwave background discovered by Penzias and Wilson in 1964. We have seen that it is the striking isotropy of this radiation that is a sign of its cosmic origin. And the theory of inflation precisely explains why this radiation has exactly the same properties throughout the whole sky. However, as we will see, inflation predicts that, at a certain (although very weak) level, anisotropies appear.

Anisotropies of the cosmic microwave background

We have seen that, before recombination, matter and light interact and form what is called a ionized plasma: photons (light) cannot escape and the Universe is opaque. Moreover, the horizon has become smaller and, in each region of the size of this horizon, the plasma evolves independently of the other regions, just as if it were contained in a box.

In this box of the size of the horizon, the plasma is subject to two opposite effects. On one hand, gravitation tends to attract matter to make it collapse onto itself. On the other hand, light acts through the pressure that photons exert on matter and tends to expand the matter. These two contradictory effects (gravitational compression and expansion by light) trigger, at the lightest perturbation, compression waves in the horizon box. In a way, these waves are very similar to the acoustic waves that can be heard inside a conch shell. In the shell, the initial perturbation is noise from outside that comes and reverberates (echoes) on the walls of the shell (Figure 5.6).

In the case of our shell/horizon, the initial perturbation comes from time immemorial. It is a relic of the first moments in the Universe, of the inflationary phase.

Let us return to the phase of inflation, triggered by the variation of quantum vacuum energy. We have seen that this energy is itself associated with fluctuations of the quantum field. These induce, according to Einstein's equations, perturbations of the geometry of space–time, i.e.

Figure 5.6 A conch as a model of horizon, within which tiny perturbations develop into a sound reproducing the surf.

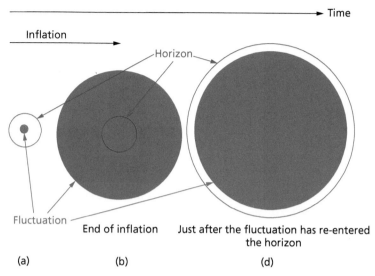

Figure 5.7 Compared evolution of the sizes of a fluctuation of the cosmic microwave background (in red) and of the cosmological horizon (in blue), during the inflation phase (a)→ (b) and well after inflation, when the fluctuation has just entered the horizon (d). See also Figure 5.4, which compares the size of the horizon with that of the observable Universe, for similarities and differences ((d) here takes place much later than (c) in Figure 5.4).

of the gravitational field. The history of these perturbations is quite interesting (Figure 5.7).

Whereas the size of the horizon remains constant during inflation, the size of these quantum perturbations increases under the effect of the very rapid expansion, until they overstep the horizon. From that moment onwards, there cannot be any dynamics associated with these perturbations (since they are too large for information, even travelling at the speed of light, to travel throughout them). But they continue to increase mechanically with the expansion of the Universe. After the end of inflation, the size of the horizon increases again. At a certain moment, the horizon catches up again with the perturbation and the quantum perturbation 're-enters' the horizon (meaning that its size becomes again smaller than the horizon).

This is why the theorists of inflation expected fluctuations in the form of anisotropies, i.e. very tiny variations of temperature through-out the sky, to appear in the CMB.

It was again the COBE satellite that gave, in 1992, the much-awaited answer: anisotropies exist at a level of 1 per 100,000 (from one point in the sky to another, variations in temperature around the average value of 2.73 K are on the order of 2.73 K/100,000). This was considered to be the first observational confirmation of the theory of inflation, a scenario that pretends to explain the dynamics of the Universe 10^{-38} s after the Big Bang!

Subsequent observations, first on the ground or in balloons, then in space with the WMAP satellite of NASA, made it possible to obtain more precise data and thus further check the theory. They have provided precise maps of anisotropies of the CMB where the sky is represented as a planisphere (Figure 5.8). The colours are chosen to identify temperature fluctuations of the CMB to some hundredth thousandth Kelvin, from the colder (blue) to the hotter (red) around the average value of 2.73 K; thus, a difference of temperature between a red zone and a blue zone is only 0.0002 K.

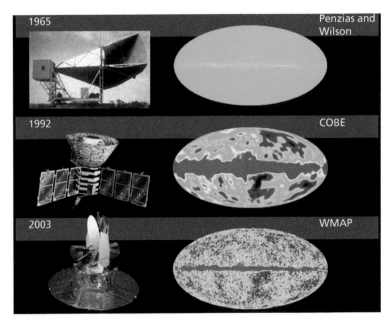

Figure 5.8 The cosmic microwave background (CMB) maps obtained by Penzias and Wilson (1965), the COBE satellite (1992), and the WMAP satellite (2003). © NASA/WMAP Science Team / NASA-JPL Caltech/ESA.

On these maps, we note the presence of a large central red band; it is due to our good old Milky Way that diffuses light in these wavelengths (microwave). To obtain a map of the CMB alone, we need to subtract this 'background noise'. The Planck mission of the European Space Agency, launched in 2009, has realized a very important work to subtract all imaginable astrophysical backgrounds and has provided maps of the cosmological background with an unprecedented level of precision (Figure 5.9).

Of course, these maps are fascinating because they provide the footprints of the first fluctuations in the Universe. But they are not only moving pictures of our past; they are also a starting point for quantitative analyses that allow us to understand the primordial Universe at the time of inflation.

We can, for example, extract from them what is called the temperature spectrum (Figure 5.10), where we identify the correlation in temperature between two points separated by a certain distance. Since we measure the cosmological background projected onto the sky, it is simpler to identify distances as viewing angles under which two points in the sky can be seen. For example, the temperatures measured at two points separated by 1° in the sky are maximally correlated (first peak on the figure), which means that there is a good chance that they are very close in magnitude. Note in this figure how the correlations seem to oscillate between successive maxima and minima.

Figure 5.9 The map of CMB anisotropies as measured by the Planck satellite and published in 2013. © ESA and the Planck Collaboration.

Understanding the oscillatory shape of the temperature spectrum

Remember our analogy of the conch. When we bring it to our ear, we hear sound waves, which are compression/dilatation waves of the air inside the conch. Similarly, for the cosmic microwave background (CMB), the original matter–light ionized plasma undergoes compression–dilatation waves inside the 'box' of their horizon (initiated by quantum fluctuations). At some point, they outgrow their horizon, reproducing themselves automatically until, at some later time, they enter again the horizon. Depending on this time, they might correspond to a succession of compressions and dilatations, be maximally or minimally correlated. It is thus not surprising that we find a peak of correlation corresponding to the time when this re-entry into the horizon occurs exactly at recombination (where CMB is produced): it corresponds to the case where the whole content of the horizon is a single giant oscillation. The size of the horizon is then 1° in the sky (Figure 5.2); it is the first peak in Figure 5.10, the most important one.

We have focused here on one of the consequences of the dynamics of the recombination phase on the photons of the CMB. But this dynamic has also left similar footprints in matter, called *baryonic acoustic oscillations*. They were successfully identified by studying the statistics of matter's distribution in the Universe (as we have just studied the statistics of the distribution of CMB photons).

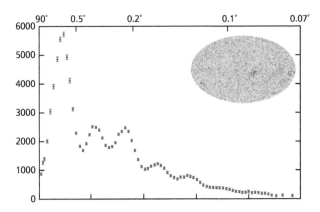

Figure 5.10 The temperature spectrum of the CMB as observed by the Planck satellite (data published in 2013). On the horizontal axis, the angle under which two points in the sky can be seen; on the vertical axis, a quantity measuring the degree of correlation in temperature.

There remains to select, among the many models of inflation, those that are consistent with all observational data, in order to construct a full-fledged theory that might give us clues about future unification between gravity and the quantum theory of fundamental interactions. This is why the data of the Planck mission and of ground detectors on the polarization of the cosmological background are so important. But to understand this, we first must explain in detail what gravitational waves are, which we will do in Chapter 8.

FOCUS V
Cosmic inflation on the ski slope

To understand the existence of this phase, we will borrow a mechanical analogy, inspired by the inclined plane of Galileo (Chapter 1).

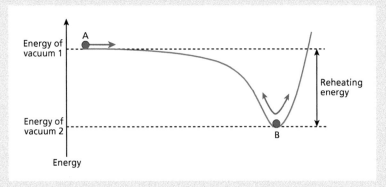

Figure V.1 A mechanical analogy of the inflation phase from vacuum 1 (marble's initial position on the left, A) to vacuum 2 (marble's position on the right, B) with the oscillation phase (reheating) where the difference of energy is consumed in heat (production of particles for the inflation).

Let us consider a board shaped as in Figure V.1: in the first part, the board is almost horizontal (but not quite); it then curves towards a region where it has the shape of a basin. A marble starts from point A with an initial velocity, which is very small. We suppose that the friction is weak but not vanishing.

The motion of the marble is easy to predict: first a very slow motion along the plateau region (this is the inflation phase which is due to the energy stored in vacuum 1), then the motion accelerates (inflation ends) before the marble falls into the basin where it will oscillate around the minimum. Because of friction, oscillations will decrease in amplitude before the marble finally stabilizes at the bottom in B.

From an energetic point of view, the excess of gravitational energy that the marble had in the initial position is transformed into heat under the effect of friction during the oscillation phase. The initial position corresponds to the initial quantum vacuum, and the position at the bottom of the basin to the final vacuum. And the energy spent in heat in the mechanical analogue is the energy necessary to produce particles in the reheating phase. Depending on the available energy, the temperature of the Universe at the end of the inflation phase will be more or less large. The end of the scenario is similar to the traditional Big Bang model where the Universe cools down little by little and becomes less dense.

6

Dark Energy and Quantum Vacuum

El vacío, la nada negada, en nuestra más vasta tradición, nuestra huella más extensa. Nuestra sombra transparente.

(The vacuum, the denied nothingness, in our vaster tradition, our more extensive track. Our transparent shadow.)

HUGO MUJICA, *El saber del no saberse* (2014)

At the end of the past century in the 1990s, a heated controversy was livening up conferences on cosmology. On one side, the theorists of fundamental physics, most of them originating from particle physics, were praising the merits of the inflation scenario, the elegant way in which it tackled and solved some of the fundamental problems of standard cosmology, such as horizon or flatness. But also the rather natural way in which this scenario appears in particle physics, its diverse realizations using scalar fields similar to the Higgs field.

On the opposite side, the community of observers, coming from astrophysics, replied that there was simply not enough matter in the Universe to obtain the critical density of 10^{-26} kg/m^3, as predicted by inflation for a spatially flat Universe. Of course, there was dark matter besides the baryonic (luminous) matter; of course, one added all forms of energy, electromagnetic radiation, neutrinos, but even inviting to the cosmic wedding all the distant cousins, one could not obtain more than 30 per cent of the critical density. Then, you know, facts against arguments based on the elegance of a theory ...

But this was not really trench warfare. The first to make a gesture were theorists who, never short of a new model, proposed the concept of open inflation, a theory of inflation that gets away with the prediction of a spatially flat Universe. But it is on the side of astrophysical observations that the decisive move came. In 1999, two teams published unexpected results on the expansion of the Universe, which implied that a major component of the Universe had been missed, not just a distant cousin but the bride's father!

Candles to map the Universe

One of the major challenges of astrophysics has always been the measurement of distances. Until now, I have largely ignored this difficulty by giving approximate distances in light years. But how to determine the exact time that light took to reach us from a given star or galaxy?

Over years has been built what we call the cosmic distance ladder. We start by measuring the distance to the Sun, then to the nearest stars by measuring the shift in their apparent position due to the revolution of the Earth around the Sun.

For more distant stars, astrophysicists have to rely on a law that relates the absolute luminosity of the star, i.e. the energy radiated by the star, to its intrinsic brightness, which is its surface temperature. This law is valid for a certain number of stars, said to be in main sequence. The surface temperature is measured from Earth by spectroscopic methods (measurement of the frequency of the light emitted by the star). We then deduce its absolute luminosity. The luminosity measured from Earth depends on this absolute luminosity, and on the distance of the star: the further the star, the weaker its apparent luminosity measured on Earth. Comparing the absolute luminosity deduced and the apparent luminosity measured, we finally obtain the distance of the star. And so on, step by step: variable stars called Cepheids were used to identify the distance of nearby galaxies, thanks to a relation discovered by Henrietta Swan Leavitt between their luminosity and their period of variability.

The difficulty of the exercise is that any error in estimating a distance at any given level of the distance ladder immediately affects the computation of astrophysical objects further away, and thus finally on the furthest distances, the cosmological distances. It is, for example, the reason why the value of the Hubble constant, which measures the expansion rate of the Universe today, has suffered important variations over time (with an estimate varying between 40 and 100 km/s/Mpc in the past 30 years): each re-evaluation of a distance within the distance ladder has led to a re-evaluation of the Hubble constant. An analysis based on the study of the cosmic microwave background seems to have now fixed this value at around 70 km/s/Mpc.

This is where we were when the singular discovery of 1999 happened.

To avoid the type of difficulties that we just mentioned, astrophysicists tried to identify a standard candle, that is, a light source that produces the same absolute luminosity, whatever the epoch of the Universe when it shines. The apparent luminosity measured from Earth makes it possible then to identify the distance. To understand the method used, imagine that you are trying to identify the size and geometry of a totally dark cave where you happen to be. If lighted candles, all identical, have been dispersed in the cave, you can identify how far they are by measuring the light received and thus reconstruct the shape of the cave. In the 1990s, a potential standard candle caught the astrophysicists' attention: the explosion of supernovae of type Ia.

The explosions of supernovae (of type II) are very violent explosions that occur at the end of the life of a massive star: the nuclear fuel has vanished and the centre of the star collapses under the sole effect of gravity, which releases a large amount of energy that makes the outer layers of the star explode. The core of a star turns into a very compact object: a neutron star or a black hole. We will come back to this in the next chapter.

If the star is less massive (at most a few solar masses), there is no explosion: the outer layers form a nebula and the core forms a white dwarf. It is this white dwarf that is of interest to us in the case of a supernova of type Ia: it is stable only if its mass is smaller than about one solar

Figure 6.1 Artist's view of the supernova explosion of type Ia, with, on the left, the exploding white dwarf and, on the right, the companion star. © NASA/CXC M. Weiss.

mass (a limit identified by the Indian astrophysicist Subrahmanyan Chandrasekhar). But many stars are in double systems. It is thus not rare that a white dwarf has a companion star (Figure 6.1). By gravitational attraction, it will little by little suck matter out of its companion and its mass increases until it exceeds the Chandrasekhar limit: its central core collapses and the heat that is thus produced makes the white dwarf explode. This is called a supernova of type Ia. The energy released in the explosion is the total energy stored in the white dwarf, which makes the luminosity depend very little on the circumstances of the explosion.

This is why, in their quest for a standard candle, astrophysicists turned to this type of violent phenomenon.

Several teams of cosmologists thus looked for these supernovae in the 1990s. The method is relatively simple: you take picture of the same section of sky at regular intervals (several weeks) and compare them. If a luminous spot appears, there is an important possibility that a star has exploded and has produced a supernova. You check with a more powerful telescope (Figure 6.2), then make an analysis of the spectrum emitted by the brand new supernova. This spectrum is characteristic of

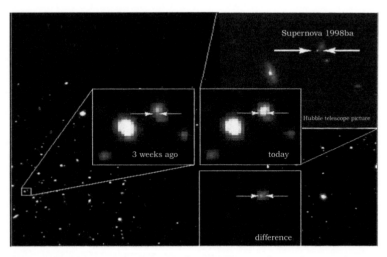

Figure 6.2 Discovery of a supernova by comparison of two pictures taken at a three-week interval from telescopes on Earth (by the Supernova Cosmology Project). The presence of the supernova is then checked on a picture of the Hubble Space Telescope. © Hubble Space Telescope NASA.

the type of supernova: hydrogen lines for supernovae of type II, but not for those of type I which, in contrast, have traces of intermediate elements such as silicium (type Ia). You thus deduce the type of supernova discovered.

But the teams working on the subject realized a rather surprising fact. Obviously, as we saw for the stars, the further away the supernovae are, the weaker their apparent luminosity is. We can use this to predict the apparent luminosity once we know the distance, assuming they are standard candles; i.e. they all have the same absolute luminosity. To obtain the distance, we measure the spectral redshift of a supernova: the evolution of the Universe, according to the Big Bang Model, predicts the Universe's expansion and thus allows us to translate the redshift into a distance (see Figure 4.2 for example). Once we know the distance, we can predict the apparent luminosity of the supernova. The surprise came when the teams realized that distant supernovae were seen from Earth not as bright as expected.

This could have two explanations: first, the hypothesis of the standard candle could be wrong; in other words, older (i.e. distant) type Ia supernovae explosions are not as powerful as younger (i.e. near) ones, they are thus intrinsically dimmer. Or these supernovae were further than expected, which means that the expansion since the explosion has been faster than the standard Big Bang Model tells us. The expansion of the Universe has accelerated!

The second possibility has consequences of utmost importance. Indeed, the known components of the Universe, matter and radiation, have a tendency to decelerate the expansion of the Universe. If expansion accelerated, it means that there exists a new component in the Universe, a new form of energy that we have been unaware of until now!

The first reactions of the scientific community were mixed. The supporters of the theory of inflation got quickly enthusiastic: here is the missing component to explain why the Universe is spatially flat. On the astrophysicists' side, the dominant attitude was rather a polite doubt. It was not the first time that an astrophysical source, presented as a standard candle, turned out to depend on the circumstances of its birth and thus on the history of the Universe.

As a matter of fact, critics went on, supernova explosion models are not even complete (to this day, no numerical simulation has reproduced the onset of the explosion): How can we be sure then that these explosions are all identical? Indeed, teams working on the data have had to partly adjust the light curves (variation of the luminosity with time) of the sources to make them look similar: more than standard candles, we should call them standardizable candles.

All this criticism was known to the research teams, who provided considerable work to find solutions: study of the supernovae environment, identification of subclasses, etc. Groundwork continued in the months and years that followed. However, what convinced the community was the confirmation of the result by other completely independent means.

One of them was provided by detailed analysis of the cosmological microwave background. Indeed, it turns out that the position of the first peak in the temperature spectrum (see Chapter 5 and Figure 5.10) is a good indicator of the total energy density. This is not surprising: remember that these peaks result from oscillations in the 'box' formed by the horizon at the time of recombination. Calculations show that the first peak corresponds to $1°$ separation in the sky if the Universe is spatially flat, that is, if the density today has precisely the value 10^{-26} kg/m^3. You can check for yourself in Figure 5.10 that this is the case. Hence data from the cosmic microwave background confirmed that the Universe is spatially flat at the time when data from supernovae made it possible to understand that an important component of the Universe had until now remained in the dark!

Other data, in particular on the formation of large structures leading to galaxy clusters, came to confirm this vision (called, at the time, concordance model). And in 2011, Saul Perlmutter, Adam Riess, and Brian P. Schmidt received the Nobel Prize in physics for the discovery of the acceleration of the expansion of the Universe thanks to the study of supernovae of type Ia.

'Dark energy', just an enticing formula?

Very soon, this new component was nicknamed dark energy. The American cosmologist Michael Turner is behind this label, which made the notion very popular.

The meaning is rather straightforward. The energetic budget of the Universe showed that all forms of energy known in the 1990s, in particular in the form of matter (mass energy), accounted for at most 30 per cent of what was necessary for a spatially flat Universe. The latter fact being confirmed, a new form of energy had to account for the remaining 70 per cent of the total budget! It could not be a known form, since all these contribute to decelerating the expansion. And it could not emit electromagnetic radiation since this would be visible, as are the stars, the galaxies, and the cosmological background. It is, thus, a form of dark energy, whose nature is for the moment unknown: the only property we are sure of is that it contributes to accelerating the expansion.

The term was immediately adopted by the scientific community, as well as the general public, for good reasons and not so good ones. With dark matter, dark energy forms 95 per cent of the content of the Universe. What is more appealing than to imagine that physicists ignore the nature of 95 per cent of the Universe? Especially a Universe where the dark forces are operating?

You can imagine that some books on dark energy ended up in the esotericism shelves of bookstores, usually more furnished than the science shelves.

The expression has played a role even in the scientific community. An experimental programme of unprecedented importance has been set up to understand the nature of what forms the major part (70 per cent) of the Universe. It would have been less attractive to present it as a way to understand why the expansion of the Universe is accelerating, an expansion not even felt at the level of our galaxy.

As any successful expression, dark energy is a superb advertising slot. But what is hiding behind it?

First of all, it could be not a new component of the Universe, because it is possible that this accelerating expansion is simply due to the fact that we do not have the right theory of gravitation. In other words, that we must correct Einstein's theory at cosmological distances. This would not be so surprising, because, after all, cosmology only recently became a quantitative science, which is a science capable of giving precise predictions at cosmological distances. Then, what about modifying Einstein's theory? Not so easy: many teams around the world have gotten to work on this. The difficulty is finding another theory that reproduces all the successful predictions of general relativity *and* explains the expansion's acceleration. Most propositions suffer from major defects: presence of instabilities, particles traveling faster than light, problems with causality, etc. This does not mean that the effort is doomed; it simply shows that general relativity is a rather unique theory and that the alternative theory searched for is probably just as unique, if it exists.

Often the introduction of a new form of energy, dark energy, is opposed to a solution that would involve modifying the theory of gravity. But the distinction is somewhat artificial, because alternate theories of gravity usually introduce new fields besides the graviton and this new form of energy. This is why we tend to talk in both cases of the dark energy problem. Let us recall that the only observational fact is the acceleration of the Universe's expansion, an acceleration that is rather recent in the Universe's history since it started at a redshift of order 1. We now have observational data at redshifts larger than 2, which indicate that the Universe's expansion was still decelerating at that time.

If thus there exists a known component of dark energy in the Universe, this component must have been dominant only recently. We have already encountered this type of situation. At the beginning of Chapter 4, we saw that radiation was the dominant form of energy in the primordial era, before matter took over just before the period of recombination. The reason is that different forms of energy have different behaviours with temperature: they are all present but, depending on the epoch, one or the other prevails. Thus, the energetic budget of the Universe has evolved: first dominated by vacuum energy (inflationary phase), then by radiation until the mass energy of matter took over, and finally dark energy.

How to characterize dark energy?

Indeed, how to characterize a component of the Universe that, if dominant, tends to accelerate the expansion? Before answering this question, let us note that we have already identified such a component: the phase of inflation, a phase of strong acceleration of the expansion, is dominated by vacuum energy. The energy of the vacuum is indeed a serious candidate for dark energy.

Let us consider more closely the acceleration of the expansion. The relevant quantity is pressure. We saw in the previous chapter what is radiation pressure. This pressure, a force per unit area, is measured in the same units as energy density ($kg.m^{-1}.s^{-2}$); effectively, in the case of electromagnetic radiation, pressure is equal to one-third of the energy density. Such a relation is characteristic of radiation and is called the equation of state of radiation. The factor ⅓ is called the equation of state parameter. Matter at rest, on the other hand, has a vanishing pressure: its equation of state parameter is 0 (i.e. its pressure is 0 times energy density). The quantum vacuum also as an equation of state: its pressure is negative, exactly opposite to its energy density; its equation of state parameter is −1! But what is a negative pressure? We will see that only components with negative pressure can accelerate the expansion of the Universe: dark energy must therefore also have a negative pressure.

Warning

Regarding dark energy, there are many misconceptions, sometimes conveyed by physicists themselves. For example, one idea is that dark energy is associated with a repulsive force that counteracts the attractive force of gravitation. The example of Einstein's static solution where the cosmological term counteracts the collapse of matter into itself has even been cited. All this is incorrect: if it were the case, then the primordial Universe, where dark energy is subdominant, would be contracting and not expanding. The misunderstanding arises from confusing the Universe with its sole matter content, and confusing notions of spatial locality (matter localized in a volume) with global temporal notions (the rate of expansion of the Universe).

We know that if molecules of gas, at a nonzero temperature (i.e. above absolute zero), are enclosed in a box, they exert outward pressure forces on the box's walls. The exerted pressure is positive (just like the radiation pressure on a radiometer). It would be negative if the pressure forces were exerted inwards. An example of the latter case is provided by the Casimir effect (Figure 6.3): two conducting plates of large dimensions, placed in the vacuum, are arranged parallel to one another at a very small distance (on the order of a few microns). They are not electrically charged; nevertheless, they are attracted to one another. The effect is due to the quantum fluctuations of the electromagnetic field between the two plates, i.e. to the particle–antiparticle pairs that keep appearing and disappearing. This effect can be computed using the quantum theory of fields and indeed precisely predicts the force measured experimentally. Compared to the preceding example, the molecules of gas have been replaced by the quantum fluctuations of the vacuum and the pressure forces act towards the interior of the cavity made by the two plates: the pressure of the vacuum is negative.

We saw earlier that, when one goes from radiation to matter at rest, the pressure goes from a positive value to a vanishing one; in parallel,

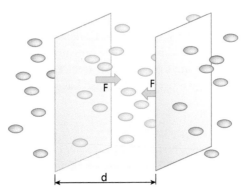

Figure 6.3 Principle of the Casimir effect. The quantum fluctuations of the electromagnetic field, due to the creations and annihilations of particle–antiparticle pairs, represented by bubbles, induce an attractive force (F) between the two conducting plates.

Why the Casimir effect is attractive

It is somewhat easy to understand why, in the Casimir effect, the force between the plates is attractive. You just have to remember that any system naturally evolves to minimize its energy. Let me now move the two plates away from one another: the cavity between the two plates becomes larger, it encompasses more vacuum, thus the vacuum energy it contains grows. Thus, the natural motion of the plates being such that they decrease the energy of the cavity, they move toward one another, they attract one another.

the deceleration of expansion decreases. We can reasonably imagine that if it goes towards sufficiently negative values of pressure, the deceleration becomes negative, in other words, there is acceleration. We are going to verify this with the test of a lift in free fall (Figure 6.4). Imagine that this lift is falling above the surface of the Sun. The pressure of solar radiation will act on the lower end of the lift and diminish its acceleration, in other words decelerate the lift. On the other hand, if we have a hypothetical source of quantum vacuum fluctuations, the motion will be accelerated by the negative pressure exerted by the fluctuations.

We conclude that dark energy must be searched for among the components that generate a sufficiently negative pressure. More precisely, theoretical applications show that the equation of state parameter, noted w, must take a value between −1 (the value it would have if dark

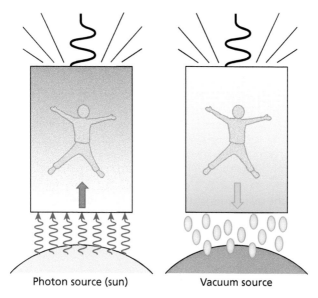

Photon source (sun)   Vacuum source

Figure 6.4 A lift in free fall, on the left, above the Sun (the photons induce a radiation pressure directed upwards), and, on the right, above a hypothetical source of vacuum (the fluctuations induce a force oriented downwards).

energy is vacuum energy) and $-\frac{1}{3}$. The observational question is then: what is the exact value of this parameter between -1 and $-\frac{1}{3}$? If it is -1, it is a strong argument in favour of vacuum energy.

Is dark energy the energy of the quantum vacuum?

Since the beginning of the year 2000, many models of dark energy have been proposed. Many are inspired by inflation models because in inflation the expansion of the Universe accelerates as well. Indeed, aren't we at the beginning of a new phase of inflation?

Most models use a scalar field (just like inflation models) with a dynamics that ensures a negative pressure. The simplest models are called quintessence, name given in ancient Greece to ether, the fifth essence that philosophers added to the four traditional elements (earth, wind, fire, and air): a kind of defiance to Einstein who showed, thanks to the result of Michelson and Morley (Chapter 1), that there is no ether supporting electromagnetic waves.

Quintessence

The quintessence field is a field that varies with time. It is a little like the inflation field at the end of the inflation period. The situation would be that of a marble when it is sufficiently far from the initial position A (in Figure V.1). Indeed, in A, the marble is almost at rest, which corresponds to the inflation phase where vacuum energy dominates: pressure is then maximally negative. When the marble start moving faster, we can imagine the pressure is less negative ($-1 < w$) but sufficient to accelerate the expansion ($w < -\frac{1}{3}$). It is what calculations confirm.

One of the difficulties of the scenario of quintessence, and of many dark energy scenarios, is that the scalar field has an extraordinarily small mass: the only constant coming into play is the Hubble constant that characterizes the expansion rate today. Expressed in energy scale, this gives a mass energy of 10^{-33} eV! Consequently, the particle associated

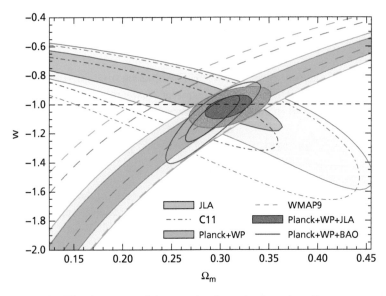

Figure 6.5 The observational data coming from the detection of supernovae (in blue) and cosmological background (in green) constraining the parameter w of the equation of state of dark energy and the ratio of matter density to critical density: the central favoured values (in grey) are w $= -1$ and a fraction of 30%.

with the dark energy field can be exchanged between two elementary particles (for example, quarks); it plays the role of the mediator of a new force, just as the photon is mediating the electromagnetic force, or the graviton the gravitational force. In the last two cases, the force has an infinite range because the mediating particle has a vanishing mass. Here, the particle is so light that the range of the force is the size of the observable Universe.

We thus have a new long-range force, but one that is not constrained as the gravitational force is, by general relativity principles. In particular, there is no reason that it satisfies the constraints imposed by the equivalence principle, although they are well tested experimentally and thus apply to any long-range force. Most dark energy models experience difficulties confronting such limits.

Figure 6.5 summarizes the constraints on the w parameter: we see that, for the moment, the favoured value is the one that corresponds to vacuum energy ($w = -1$). An ambitious experimental problem has been set up to obtain a more precise determination of this parameter, and confirm or infirm this conclusion.

Existential problems

It thus appears that the most probable candidate for dark energy is vacuum energy. If this were confirmed, the problem of vacuum energy and of its computation would become our main concern. But this calculation requires interfacing general relativity with quantum theory, a task whose utmost importance we have already underlined. For the record, nongravitational physics can only measure differences in energy: for example, the measurement of the Casimir effect goes through the measurement of a force, which can be seen as resulting from the energy difference when the distance between the plates is changed (Figure 6.3); this effect does not measure the absolute value of the energy of the quantum vacuum. But in a gravitational context, according to general relativity, any form of energy participates in the Universe's expansion, and is thus measurable in principle. It is, in particular, the case for vacuum energy: if it is dark energy, it represents 70 per cent of the critical density, i.e. its energy density is 0.7×10^{-26} kg/m^3, a quantity absolutely measurable!

Do we know how to compute it? In order to do so, we would need a theory of quantum gravity. We are somewhat in the position of Galileo trying to elucidate the laws of classical gravity (see 'Galileo and Dimensional Analysis' in Chapter 4): physicists thus follow his example and use dimensional analysis. The vacuum energy density should be expressed in terms of the scale of quantum gravity: the Planck scale. The value obtained is 10^{94} kg/m^3, i.e. 120 orders of magnitude larger than what is measured. This gives an idea of the gigantic gap that exists today between general relativity and quantum physics: by using the theories currently available, we are wrong by 120 orders of magnitude, a gap that we must absolutely fill if it is confirmed that dark energy is vacuum energy. The problem is so huge that we can expect that once we have identified the theory that unifies the quantum and gravitational approaches, this solution will be obvious.

What is the situation with the theories currently studied? The least we can say is that they are not very comfortable with this altogether fundamental question.

In particular, string theory, which we talked about in Focus IV in our discussion about the Big Bang, has had to resort to the anthropic principle to solve the question. According to this principle, constants of nature have the value they have in order to allow for the existence today of observers that can measure them. This can be understood in the context of multiple universes, or multiverses, evolving in parallel: we then talk of the probability that we are in one or another universe. And it is clear that this probability is vanishing in all universes where fundamental constants have values such that we cannot exist today and observe the sky.

The argument is more or less as follows: if the energy of the vacuum had been much larger, the phase of acceleration would have started much earlier and would have prevented the formation of lumps of dark matter that gave birth later to galaxies. Among the universes compatible with the existence of galaxies, it is more probable that we are in a universe where the phase of acceleration has already started, because this Universe occupies a larger volume than a universe where the vacuum energy would be much smaller and would have not yet started dominating. In other words, the vacuum energy can be never much smaller nor much bigger than what is observed. Needless to say, the debate is raging within the scientific community to decide whether this is a new way of making scientific predictions, or just a cruel admission of helplessness.

How to conjure up a mental picture of the quantum vacuum?

Infini rien.
(*Infinite nothing.*)
BLAISE PASCAL, *Pensées* (1670)

Tao produces Unity,
Unity produces Duality
Duality produces Trinity
Trinity produces the ten thousand things
The ten thousand things are sustained by Yin
They are encompassed by Yang
And they are harmonized by the breath of Vaccuum.

LAOZI, *XLII*

The notion of quantum vacuum is difficult to grasp because it involves the infinitely small (where the laws quantum physics are valid) and the infinitely large (the whole Universe). How do physicists imagine it? How should it be represented? Because art is both representation and interpretation, artists sometimes establish short circuits that allow us to come closer to these difficult notions.

For the Westerner, the word vacuum is in itself an obstacle to understanding the concept. Indeed, since the vacuum is what remains where and when everything has been taken away, how is it that physicists see in it a structure, symmetries, fluctuations, energy, etc. and that we can even evolve from one vacuum into another? This is why quantum physicists prefer the notion of the fundamental state: every quantum system is described by a collection of states built out of the state of minimum energy, called fundamental state. A state with one particle is built on the fundamental state and has an energy equal to the energy of this fundamental state plus the mass energy of the particle mc^2. A state with two particles .... We thus construct a tower of states above the fundamental state (Figure VI.1).

Then why keep the term quantum vacuum? Well, because the notion is much richer in non-Western cultures. In particular in the Chinese culture, the concept is much more elaborate and closer to what physicists mean by a vacuum. I cite François Cheng (in *Empty and Full: The Language of Chinese Painting*):

In the Chinese perspective, the Vacuum is not, as one could suppose, something vague and non-existent, but an element it is eminently dynamical and acting. Related to the idea of breath of life and of the principle of alternation Yin–Yang, it constitutes par excellence the scene where transformations operate, where the Full would be able to reach the true fullness.

We see that, in Oriental culture, the notion of vacuum is much more dynamic than for Westerners, and is engaged in a fruitful dialogue with the notion of fullness. In the same way, in quantum physics, the full is constructed on the vacuum. The parallel is actually striking between the famous citation of Laozi reproduced in the epigraph of this Focus and the quantum construction of Figure VI.1.

Another image that comes to a physicist's mind when he is conjuring up the quantum vacuum is a swarm of quantum fluctuations. Particle–antiparticle pairs are generated from the vacuum and then

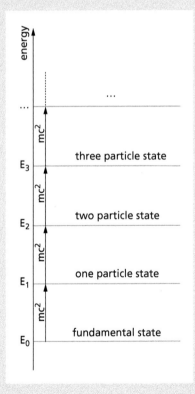

Figure VI.1 Tower of quantum states built on the vacuum (fundamental state).

recombine to turn back into vacuum. This is a violation of the conservation of energy of twice mc^2 (particle and antiparticle have the same mass m), but quantum mechanics allows this type of violation during a time all the shorter that the mass is larger. Such particles are called virtual because they cannot be detected: otherwise, violation of energy would be blatant. This means that the quantum vacuum is virtually rich in all particles, in all potentialities of the Universe in a sense. And it is this potentiality that yields energy to the vacuum.

How can the quantum vacuum be represented? The first example that comes to mind is a Japanese Zen garden, more precisely the expanse of loose gravel combed by the rake of the gardener, which forms the background of a Zen garden on which stands out a few arrangements of stones or trees. This background is as undefinable as the quantum vacuum is, but on it is constructed the garden-Universe, enclosed by a wall that marks a limit, but in a permanent dialogue with the exterior world, a blooming tree or a foliage that tops the wall.

In great historical Zen gardens, such as the Ryoan-ji (Figure VI.2), there is also a temporal dimension, symbolized by the wooden platform that looks down on the garden: generations of visitors who have come to contemplate the garden and meditate have bestowed on it their thoughts, their dreams, their experience of life. The garden is rich in all these virtual stories, all these potentialities, and has become over centuries much more than an expanse of gravel with a few standing stones. A sort of quantum vacuum ...

I remember a concert in the Andoain quarry in the Basque country. The electro-acoustic music of the composer Gorka Alda, entitled *Fluctuations of the Vacuum*, was filling the void of the quarry, recreating the virtuality of a mountain, of the mountain that had disappeared stone by stone during the quarry's operation. Each stone was reconstructed by a sound fluctuation.

Another image: in our laboratory in Paris, the sculpture *Squaring the Circle* by the artist Attila Csorgo is hidden in a case of darkness (Figure VI.3). Light reflected by a clover-shaped mirror projects on the ground the square shadow of a circular disk. The circle is a Universe of its own and the piece reveals that it has within itself the potentiality of being square: the circle as a virtual square. What is then the quantum vacuum in this work of art? To me, it is the light because it is the medium that reveals this potentiality.

Figure VI.2 The Japanese garden of Ryoan-ji. © PBGA.

How can the quantum vacuum be represented? This was the question asked of a few scientists during LabOrigins, a workshop organized by the journalist and science facilitator Marie-Odile Monchicourt with a group of scientists and artists, to prepare a public performance focused precisely on the quantum vacuum. I must confess that, during these sessions, we physicists had great difficulty characterizing the quantum vacuum, and extracting ourselves from our predefined schemes to give a vision of this vacuum. Meanwhile, through the rehearsals, the artists had started to focus: a metallic structure appeared on the stage (had we talked of the structure of the vacuum?), all in a breakdown of symmetries. And the dancers-acrobats took ownership of this structure, expressing with their gestures the domination of gravity. Had we insisted on the central role of vacuum energy in gravity? I do not recall …

The coherence of the artistic approach made these striking shortcuts possible, where a connection that we scientists have slowly worked out is naturally established in an intuition of the emotion or simply of the body.

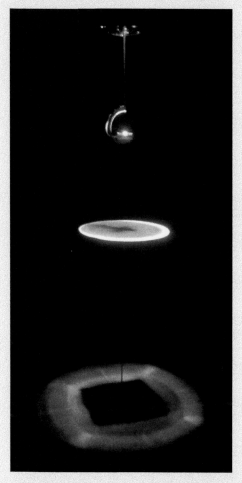

Figure VI.3 Squaring the Circle, by Attila Csorgo. © JLR-APC.

Indeed, the vacuum, as a form of energy, contributes to gravitation. Some years ago, I wrote a small text entitled 'The Unbearable Lightness of the Vacuum'[1] to express the fact that the vacuum is much lighter than it seems. This is what we physicists soberly call the vacuum energy problem. Will understanding gravity and how it fits into the laws of the quantum world allow us to solve this problem? Maybe artists should start thinking about the question?

---

[1] As a small tribute to the book *The Unbearable Lightness of Being* by Milan Kundera.

7

Leçon de ténèbres:
Black Holes

The undiscovered country from whose bourn,
No traveller returns, puzzles the will,

WILLIAM SHAKESPEARE, *Hamlet* (1601)

Black holes are often presented as a direct consequence of Einstein's theory, which is exact. But their existence itself prevailed only very gradually among scientists, and Einstein himself was never convinced of it.

In this long history, a little revolution occurred at the beginning of the twenty-first century that passed almost unnoticed. Black holes evolved from the status of *putative* celestial objects to that of *fully fledged* astrophysical objects, and even rather common ones. In short, this status changed from hypothetical black holes to astrophysical black holes within only a few years. What were the reasons for this change of attitude? It was the conviction that the centre of our own Galaxy is the home of a black hole, whose mass would be equivalent to the mass of a few million Suns! This realization completely changed points of view, and all previous indications of black holes' presence in the Universe (quotation marks were used when referring to them) turned into confirmations of their existence.

Many of the most intriguing aspects of the theory of gravitation are concentrated in the physics of black holes. This is why it is probable that these fascinating objects will play a central role in solving the fundamental questions we still have about gravity, in particular on its connection with quantum physics.

The concept of a black star dates back to the eighteenth century. We mentioned in the first chapter Newton's cannonball experiment: by giving a material body placed on the Earth's surface a velocity of 7,900 m/s, we can set it into orbit. Giving it a slightly larger velocity will allow it to lift away from the gravitational attraction of the Earth: this is called the escape velocity, whose value on Earth is 11,200 m/s, *independently of the mass of the body.* If we perform the same experiment on a different planet or star in the Universe, the more massive the celestial object is, the larger the escape velocity will need to be.

In 1783, John Michell imagined the limit case of a star so massive that the escape velocity would be larger than the speed of light. His reasoning followed Newton's interpretation of light made of corpuscles: these grains of light cannot escape because their velocity is smaller than the escape velocity. Light emitted from the star is trapped, just as light that falls onto the star is. This star absorbs any form of light and is thus a black star (the term 'black hole' only appeared later). The idea was rediscovered in 1796 by Pierre-Simon de Laplace but was considered then as a mathematical artifice. No such object could possibly exist in nature!

The next chapter was written in 1915. We saw in Chapter 2 that Karl Schwarzschild obtained a solution of Einstein's equations that is valid in the vacuum, *outside* a spherical star. But the solution has a singular mathematical behaviour in two zones: at the exact centre of the star and on a sphere of a specific radius, known as the Schwarzschild radius, whose exact value depends only on the mass of the star and on fundamental constants. But the Schwarzschild radius has a tiny value (3 km

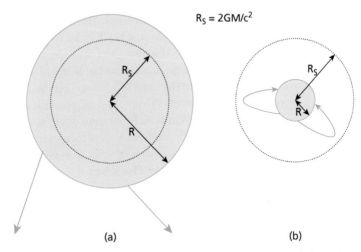

Figure 7.1 Schematic behaviour of light emitted (green arrows) from the surface of a star of radius R (a) when the Schwarzschild radius R_S is smaller than R, and (b) when it is larger. The Schwarzschild solution is only valid in the vacuum, i.e. outside the star.

for the Sun of mass 2×10^{30} kg) so these two regions are deep inside the star, where there is matter (Figure 7.1a): the Schwarzschild solution, only valid in the vacuum, is thus not applicable.

But what happens if the star is so massive and compact that its own radius is smaller than the Schwarzschild radius (Figure 7.1b)? Then the star behaves like the black star of Michell and Laplace. For a long time, this seemed just science fiction: compare the radius of our Sun, 700,000 km, with the value of its Schwarzschild radius mentioned above!

But the story took another twist with the work of Subrahmanyan Chandrasekhar, and then Robert Oppenheimer and Hartland Snyder on the final evolution of black holes.

The death of a star

The death of a movie star sometimes turns them into a legend. We might say the same of astrophysical stars, but black holes are no legend!

I mentioned briefly in the previous chapter the final evolution of a star. Stellar dynamics is the result of two contradictory effects: gravitational attraction that tends to contract the star into a smaller volume, and nuclear reactions that produce heat that tends to dilate the star. In a star like our Sun, these two effects cancel: this is why its size remains constant. We have also seen that nuclear reactions play an important role: they synthetize heavy elements from the hydrogen and helium made in the primordial eras. Via gravity, these heavier elements fall to the core of the star. Once the nuclear fuel is burnt out, the balance between the two effects is broken: gravitational attraction comes to dominate and induces the collapse of the star core into itself.

Can the star avoid turning into a black hole? Is the collapse stopped by some physical process? Indeed, and this is where quantum physics enters the picture. Particles that form ordinary matter (protons, neutrons, electrons, etc.) obey the Pauli principle: they cannot be in the same quantum microscopic state. As we saw in Chapter 4, they are fermions ($1 + 1 = 2$). When matter undergoes gravitational collapse, the process is slowed down by the resistance of elementary fermions to interpenetration. Imagine trying to crush a Lego construction: you may destroy it but you would have to exert an enormous pressure to embed one Lego brick into the next one. In quantum physics, this is called degeneracy pressure.

If the star is not very massive (more precisely if its mass is smaller than a limit value on the order of 1.4 solar mass, called the Chandrasekhar limit), the collapse is stopped and the core of the star becomes a *white dwarf*. It is the degeneracy pressure of electrons that stops the downfall. The complete process creates some heat that dilates the outer layers: the star appears as a red giant, before transforming into a nebula, with a white dwarf in its centre (Figure 7.2).

If the star is more massive, the degeneracy pressure of electrons is insufficient to stop the collapse, and electrons and protons combine into neutrons (and neutrinos that escape). It is the degeneracy pressure of neutrons that then stops the downfall. The core of the star becomes a *neutron star*, a very compact object: the mass is between 1 and 3 solar masses and the radius is on the order of 10 km.

If the star is even more massive, then nothing can prevent the gravitational collapse: matter falls all the way to the centre, and a *black hole* forms.

We see that, in order to fully understand the dynamics, a complete version of quantum gravity is needed. As a matter of fact, the singular

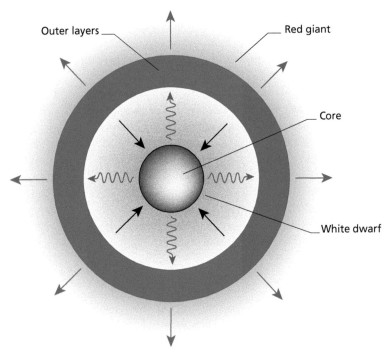

Figure 7.2 Formation of a red giant. The gravitational collapse of the core generates thermal energy that dilates the outer layers of the star.

behaviour at the centre of the black hole, where density should become infinite, appears to have strong similarities with the Big Bang singularity. But there is a fundamental difference: if we send an observer, say a mere photon, to check what is going on there, it must reach a distance from the centre that is smaller than the Schwarzschild radius, and will thus be unable to come back and report to us its findings. In some sense, the central singularity of the black hole is isolated from us in a fundamental way, in contrast to the Big Bang singularity, the observation of which we can imagine experimental ways of conducting.

The spherical surface around the black hole, with a Schwarzschild radius, is called the black hole horizon. We have seen that the original Schwarzschild solution had a singular behaviour precisely at this distance, but subsequent studies showed that this was only due to the coordinates chosen by Schwarzschild, and was not of a fundamental nature, in contrast to what happens at the centre of the black hole. It remains that the horizon surface is a very peculiar place; it is the surface of no return: every object that goes through it, even a photon, is doomed to fall into the central singularity. On the other hand, an object that gravitates around the black hole at distances larger than the Schwarzschild radius may orbit it in a standard way: contrary to a misconception, a black hole does not swallow all mass and light but only those that come close to the horizon.

A simple example will enable you to understand what is the horizon of a black hole, and what it is not: the one of a waterfall (Figure 7.3).

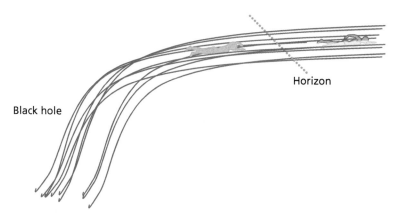

Figure 7.3 Analogue of a black hole horizon using a waterfall.

We have all swum in a river without worrying about the presence of a waterfall down the river. However, there exists near the waterfall a line with quite real consequences: if we cross it when following the stream, then, even if we are champion swimmers, we are irredeemably pulled along by the current into the 'singularity' of the waterfall. This fictitious line that is not identified in the river is the horizon of the waterfall.

In the same way, the observer that is falling towards a black hole's horizon receives no warning call when he actually crosses it. Nevertheless, he is lost forever, at least for those of us who stay outside. And his messages, coded into electromagnetic waves, will indefinitely stay on the other side.

The name 'black hole' was only proposed in 1967 by John Wheeler. A few years earlier, in 1963, another solution of Einstein's equations had been obtained by the New Zealand mathematician Roy Kerr that described rotating black holes. We will see that most black holes in the Universe are probably Kerr black holes, i.e. are in rotation.

The first candidate black hole that was clearly identified was observed with X-rays in 1971. It is one of the two stars of the Cygnus X-1 binary system: matter torn away from the companion star (the other star of the binary system) falls into the black hole horizon while emitting X-rays (Figure 7.4). It was verified later that the compact object has a mass around 6 solar masses, which is too massive to be a neutron star and confirmed its alleged status as a 'black hole'.

Figure 7.4 The Cygnus X-1 binary system: on the left, position in the sky; on the right, artist's view with the black hole (disk and jets) progressively sucking the matter from the companion star. © NASA/CXC-Digitized Sky Survey.

The years 1960 and 1970 were also prolific years for the theoretical understanding of black holes, for the works of Brandon Carter, Stephen Hawking, and Roger Penrose in particular. We will return to this later.

Black holes and galaxies

Candidate black holes multiplied in the 1970s, with more or less convincing arguments of their existence. But the discovery of a massive black hole in our own Galaxy managed to overcome the psychological barrier that stood between candidate black holes and full-fledged black holes.

The central region of our Galaxy, which lies some 25,000 light-years from us, is the home of strong activity in several frequency ranges of electromagnetic waves (gamma, infrared, and radio). In particular, an astronomical radio source, called Sagittarius A*, appears to be at the geometric centre. Because the centre of many galaxies seems to be home to a very massive black hole, the hypothesis is that the emission of radio waves is due to the accretion of gas onto a black hole hidden at the centre of Sagittarius A*.

This was spectacularly confirmed in 2003 by the infrared observation of the motion of stars very close to Sagittarius A* (Figure 7.5). These stars,

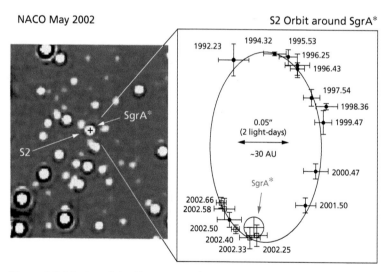

Figure 7.5 Motion of the S2 star around Sagittarius A* at the centre of the Milky Way. The different positions are labelled by the year of observation (from 1992 to 2002). © ESO https://www.eso.org/public/images/eso0226c/

Figure 7.6 Image of the central zone of the NGC 4261 galaxy with two opposite jets projecting matter as far as 90,000 light years and, at the very centre, an accretion disk of 400 light years and the surrounding dust torus. © Hubble telescope ESA/NASA.

such as the S2 star identified in the picture, orbit around the galactic centre with a period of around 10 years. Progresses in adaptive optics have made it possible to identify with great precision their successive positions and to reconstruct their trajectory. These trajectories are compatible with the gravitational attraction of an object of some 4 million solar masses. This mass should be that of an extraordinarily compact object since some of the stars get as close as one astronomical unit, i.e. the Earth–Sun distance, from its centre. You may find this a large distance but remember that the object has the mass of several million suns! In fact, the object is so massive and compact that it seems only compatible with it being a very massive black hole. A huge black hole at the very centre of our own Galaxy!

Busy black holes

If you imagine a black hole as a catch-it-all device acting as a vacuum cleaner in its vicinity, you are far from reality. In fact, a black hole organizes matter around itself in a very structured way. Figure 7.6 shows the central structure of the NGC 4261 galaxy, located some 100 million light years from us. We think that its centre is occupied by a Kerr black hole of some 400 million solar masses. Since the black hole is rotating, the axis of this rotation defines a specific direction: two opposite very energetic jets of particles can be seen along this direction. As for the matter orbiting close to the black hole, it is organized in a disk, called an accretion disk, surrounded by a torus of dust. The disk is in the plane orthogonal to the axis of rotation. The black hole horizon is at the centre of the disk and is not visible: any light that crosses it is lost forever.

This general pattern that we identify in the immediate vicinity of a massive black hole (but still outside its horizon!) is found in very diverse situations, for example around the much lighter black holes formed at the end of the life of a star. Figure 7.7 shows similar organization of matter in the case of (a) a quasar, i.e. a very massive black hole at the centre of a galaxy, fed by galactic matter; (b) a microquasar, i.e. a stellar black hole scavenging the matter of a companion star; and (c) the gravitational collapse of a massive star's core into a black hole. Mass scales are very

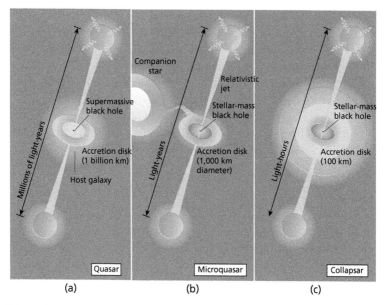

Figure 7.7 Organization of matter (jets, accretion disk, dust torus) around galactic (quasar, a) or stellar (microquasar, b; collapsar, c) black holes.

different, distance scales are very different, but the general picture is the same.

You are probably surprised to see jets of particles present: the black hole seems to eject matter, which appears in contradiction with the familiar image of a matter-eating ogre. We attribute this behaviour to the electromagnetic force. When the matter of the star collapses into a black hole, the magnetic fields reorganize themselves around the forming, rotating black hole. Part of the matter is focused into these jets that reject the matter further away. Some think that the jets start very close outside the horizon. However, numerical simulations that reproduce via computers the complex phenomena that give birth to these jets have some difficulty generating them faithfully.

If one of these jets of energetic particles, which emit light, is oriented towards Earth, then it is seen as a luminous point. This is how we are able to detect, for example, the gravitational collapse of a massive star (right panel of Figure 7.7): the light coming from the jet is known as a gamma ray burst. These bursts can be bright enough to be seen with the naked eye from Earth, but their luminosity decreases rapidly with time. They were discovered in 1967 by the US Vela satellite, which was supposed to observe the Soviet territory to monitor their compliance with the 1963 Partial Test Ban Treaty on atmospheric nuclear explosions. The discovery was only publicly announced in 1973, leaving ample time to check that little green Soviet men were not exploding bombs far from Earth!

These energetic jets have also made it possible to identify black holes hiding at the centre of galaxies. The presence of these jets at the centre of a large number of galaxies has indeed been observed (see, for example, Figure 7.6). The astrophysical sources at the origin of these jets were called *active galactic nuclei*. We now think that all these active galactic nuclei are, in fact, massive black holes. The black hole at the centre of our own Milky Way is somewhat not very active: this is why we have not detected jets. Nevertheless, we have detected past activity of this black hole in the molecular clouds that surround the galactic centre.

The central black hole probably played an important role in the dynamics of the Galaxy. Its mass is related with the total mass and size of the Galaxy. We think that the first galaxies, less structured, had a central black hole that was not so massive (10,000 to 100,000 solar masses). We said earlier that galaxies structured themselves with time by merging (Figure III.2). We think that their respective black holes also merged, giving birth to an even more massive black hole: an exceptional event to which I will return in Chapter 10. Galactic black holes have thus grown in mass and size all throughout the history of the Universe through mergers and accretion of matter. Their history is intimately correlated with the history of the host galaxy. Recent results suggest that the deficit of matter observed in galaxies might be due to the energetic jets that could expel part of the matter out of the galaxy.

Laboratories of the gravitational Universe

The generic character of the phenomena that take place in the immediate vicinity of black holes, summarized in Figure 7.7, is probably due to the relatively simple structure of black holes themselves. Quintessential gravitational objects, they appear to be characterized by only three numbers: their mass, their angular momentum (i.e. how fast they rotate), and their electric charge. Strangely enough, this likens them to elementary particles, identified by their mass, their intrinsic angular momentum, called spin, and their electric charge.

This property is a consequence of a theorem of general relativity, known as the 'no hair' theorem. The name is due to the American physicist John A. Wheeler, who is said to have exclaimed: 'Black holes have no hair!'. This comparison of a black hole to a bald head gave its name to the theorem and made it famous, even if, strictly speaking, a black hole has three hairs (the three characteristics that are mass, charge, and angular momentum). It remains to be determined whether astrophysical black holes are as simple entities as predicted by Einstein's equations. I will come back to this in the next chapters.

Because they are essentially gravitational objects (the no hair theorem suggests that they lost all original matter characteristics in the gravitational collapse that gave birth to them), black holes are superb laboratories for testing the gravitational force. In particular, because they are very massive and compact, the gravitational field near their horizon is intense: they, thus, make it possible to observe gravity's behaviour in the strong field regime and to verify that it is adequately described by general relativity. This requires experimenting close to the black hole horizon, which is impossible directly. But astrophysical objects, such as stars, commonly fall into the horizon of black holes, emitting electromagnetic or gravitational waves (we will return to them in the next chapter). This could give us more or less direct knowledge of the phenomenon at play, and provide unique information on gravity's behaviour.

Information, in the usual sense, can be decomposed into elementary bits, such as the 0 and 1 of a computer. As far as we are concerned, information can be the position and velocity of an elementary particle, together with its mass, charge, and spin. A black hole is a remarkable information safety box: every piece of information that crosses the horizon is stored. We can, thus, imagine that this information covers the black hole horizon.

To understand what I just said, imagine a safe into which money is inserted through a slit. All the information contained in the safe (total amount, number of notes of a given value, etc.) is stored in the chip that controls the slit. Imagine that we divide the horizon of a black hole into elementary surfaces (Figure 7.8) and that we equip each individual cell with a chip that registers what goes through the surface. The complete information contained in the black hole is registered in the chips that cover the horizon surface. This surface is often compared to a hologram, a 2-dimensional surface that contains the full 3-dimensional information describing a volume.

Stephen Hawking showed that we can define the quantity of information stored in the black hole as being proportional to the surface of its horizon. This quantity, called entropy, can only increase with time. Indeed, let us imagine throwing an object of mass m into a black hole of mass M. The black hole mass increases to $M + m$; its Schwarzschild radius, which is proportional to the mass, increases accordingly, and thus the surface of the horizon: entropy as well because extra information has entered the black hole. Since the inverse process (a black hole spitting out an object of mass m) is not possible, entropy can never decrease. Therefore, we have, for the black hole, the equivalent of the second principle of thermodynamics.

Stephen Hawking has also identified the phenomenon of black hole evaporation. Contrary to what we have said until now, black holes may lose energy, emit radiation, and disappear. The process behind this behaviour is of a quantum nature: we are back at the crossroads between quantum physics and gravitation.

Let us imagine a quantum fluctuation, i.e. a particle–antiparticle pair, emitted by the vacuum at the immediate vicinity of the horizon (Figure 7.9). Far from the black hole horizon, this pair would exist only for a brief moment before annihilating back into the vacuum. But very close to the horizon, it is possible that one element of the pair, say the antiparticle, vanishes behind the horizon; from then on, it is impossible to recombine with the particle. The orphan particle thus goes on its way

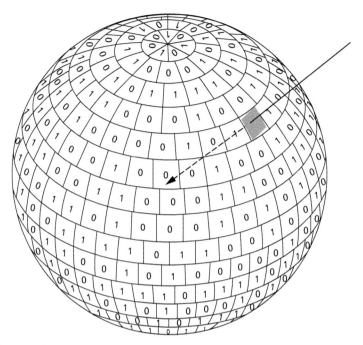

Figure 7.8 Black hole horizon surface, covered with elementary cells that register the information that went through (necessarily from outside to inside).

Entropy and second principle

The notion of quantity of information is described in physics by what is called entropy, which quantifies the degree of disorder of a system. This is easy to understand: imagine a well-kept library; a few bits of information will allow you to localize the book you want, whereas after an earthquake has turned the place into a mess, the information you need will be so much larger. According to the second principle of thermodynamics, the entropy of isolated systems cannot decrease (if you do not reorder the library ravaged by the earthquake, the books will not spontaneously find their place back on the shelves). Ludwig Boltzmann showed that the notion of entropy is directly related to the number of microscopic states, or configurations, of the system. This is not so different from what I called elementary information.

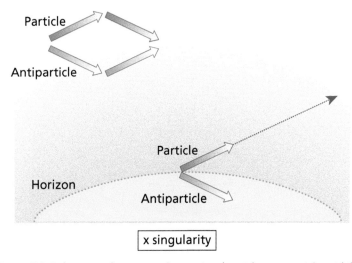

Figure 7.9 Behaviour of an energy fluctuation (particle–antiparticle pair) far (upper left) or near (bottom right) the horizon of a black hole. In the latter case, the pair is broken.

indefinitely. From the point of view of an outside observer, the black hole appears to have spontaneously emitted a particle. And indeed, it has lost an energy equal to the mass energy $E = mc^2$ of the particle (the partner antiparticle inside the black hole can be seen as a particle of negative energy $E = -mc^2$: the black hole has absorbed this negative energy and its energy has thus decreased). This is called Hawking radiation. In the long term, an isolated black hole (i.e. one that does not absorb neighbouring matter) thus sees its energy decreasing, presumably to 0, in which case it disappears. This is the phenomenon called black hole evaporation, by analogy with the evaporation of dewdrops.

This phenomenon is difficult to observe with astrophysical black holes: we are drawn to it, for example, to explain why primordial black holes, that is black holes formed during the Planck era just after the Big Bang, are not observed yet.

Some people have talked about the possibility of creating microscopic black holes when the LHC collider at CERN is turned on. This possibility was considered in the context of theories with more space dimensions. A possibility that worried some .... But, a microscopic black hole would evaporate in a microscopic time, giving rise to a puff of particles. This would not have been the end of the world, just an excellent occasion for measuring Hawking radiation in the lab!

Falling into the horizon

O God, I could be bounded in a nutshell and count myself a king
of infinite space,
were it not that I have bad dreams.

WILLIAM SHAKESPEARE, *Hamlet* (1601)

The notion of horizon, which has been applied to the black hole case in
this chapter, is inherent to any gravitational system. It can be the horizon
associated with the limit of our observations (cosmological horizon), or
the horizon determining the no-return point near a black hole. The
astrophysics of the gravitational Universe thus develops this notion,
whose epistemology is already very rich.

For the Greeks, the horizon was this circular line apparently dividing
Earth from the celestial sphere. Only apparently, since they knew Earth
was round and that remote objects were not visible because of this.
And, of course, that the horizon moves along with the observer.

The concept is more meaningful for a nation of sailors than for
mountain people. It has been enhanced along with explorations and
observations. Originally referring to a purely spatial notion, it took on
a temporal meaning that our current language abuses of: 'horizon
2020', 'a ten-year horizon', …. For a sailor of the great expeditions of the
fifteenth and sixteenth centuries, the horizon looked static—'until some
land loomed on the horizon', meaning that the ship had crossed huge
distances, and that time had been flowing.

These two characteristics—a close association with the observer and
a spatiotemporal character—remain specific features of the gravitational horizon. But the notion of horizon has many more ramifica-
tions. In a sense, the horizon separates the known from the unknown,
or from the 'not yet known'. It is the *frontier* of the American pioneer.

The horizon also makes it possible, in some way, to reconcile with
the notion of infinity. During ancient times, it represented the boundary with the perfect world of celestial spheres. It obviously contains a
finite region inside a possibly infinite space. And this is its great interest:
as long as we stay on this side of the horizon, the question of finiteness
or infiniteness does not need to be solved. Similarly, we know that
there is a horizon on Earth because Earth is round. But the horizon

prevents us from having to ask questions about this roundness: for instance, I will never be able to observe from Paris New Zealanders walking upside down!

The cosmological horizon also sets up some kind of finiteness in an infinite Universe. And it does so in a way that can be conceived even in the framework of a finite Universe (Earth's surface, for instance). In other words, the horizon is a response to our natural difficulty at grasping the notion of infinity: the horizon is, in some sense, an edge that is not an end.

There are several definitions of horizon. In this way, the event horizon demarcates the part of space–time that we will never be able to explore. In order to understand this, take a look at Figure 5.1 and imagine that you are an observer going forward on the time axis up to infinity in the future: the region from which information did not have time to reach us since the Big Bang is limited by the event horizon, a kind of future horizon.

It is not easy to visualize this infinity in the future. Don't worry: it is the same for us as physicists. But we have some tricks to get around the difficulty. One of the most popular was proposed by the British mathematician and physicist Roger Penrose. It 'simply' amounts to finding a mathematical transformation which brings infinity back to a finite distance: it then becomes possible to draw an infinite Universe on a simple sheet of paper. Figure VII.1 gives two examples of Penrose diagrams, one for a perpetually inflating Universe (called the de Sitter Universe) (a), the other for a spatially flat and matter-dominated homogeneous Universe (b). The infinite de Sitter Universe can be depicted as a square (its edges represent the points at infinity, in time or space), and the matter-dominated Universe as a triangle. In the first case, the future horizon delimitates only half of the Universe while, in the second case, it has no proper existence for it coincides with infinity: the whole space–time will eventually be accessible to our observation.

Note that the horizon spreads out in both space and time. The difficulty with the event horizon is that we need, in principle, to wait for an infinite time in the future in order to define it completely, which is subject to caution (what if the Universe evolved in a different way than what we expect?). Hence, we also define the past horizon (called quite inexplicably the particle horizon), as well as the apparent horizon.

The past horizon delimitates the region of the Universe from which we actually received information since the Big Bang (or since time immemorial if there is no Big Bang). It is shown in dots for the two Universes of Figure VII.1.

As for the notion of the apparent horizon, it can be understood from the concept of expansion. We have seen that, according to Hubble's law, the recession velocity increases with distance. Are there any points receding from us faster than the speed of light? Yes. Is this in contradiction with the laws of special relativity? No, since these points are receding from us due to the expansion of the fabric of space–time itself (as a global effect), whereas the laws of special relativity apply to the relative velocity of two objects which is measured at a given point of space–time (local effect). The apparent horizon delimitates the region of superluminal expansion. Note that we may receive information from these points, since the horizon evolves dynamically as the Universe evolves itself (Figure VII.1b, where this horizon is shown in dashed-dotted lines). Hence, some regions of space beyond the apparent horizon at a given time (their recession velocity from us is greater than the speed of light) can enter the inner zone later (where the recession velocity is below the speed of light).

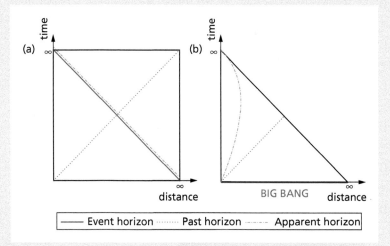

Figure VII.1 Penrose diagrams (in bold lines) for the de Sitter Universe (a) and a flat Universe dominated by matter (b). In the first diagram, the event horizon (straight line), past horizon (dotted line), and apparent horizon (dashed-dotted line) exist. In the second diagram, only the past horizon (dotted line) and apparent horizon (dashed-dotted line) exist.

In order for you to understand this, I will use a beautiful comparison, often called upon when dealing with the apparent horizon. Imagine a succession of a large number of doors. You are running through this succession of doors. In the far distance, your horizon is folding in and approaching, because the doors are gradually closing in the opposite direction of your course. At each moment a door which is closing in the distance represents your apparent horizon. After a while, a closed door will stop your run: you have reached your event horizon. With this example, you can note that at each moment you know where your apparent horizon is, that it is permanently changing (it is dynamical), but you will have to reach the end of your run to identify the door corresponding to the event horizon (you could a priori predict the position, but it depends on imponderables—such as tiredness and obstacles—which could modify your speed).

This description of the various cosmological horizons shows the central role of the observer. His role is probably less obvious in the black hole case. Yet, it is just as important. The observer can detect from very far away the Hawking radiation emitted by the hole. He can also hang slightly over the horizon: the closer he is, the more energetic the phenomenon he will be able to see. As concerns for an observer falling into the black hole ..., his fate is subject to an intense debate. We thought, until recently, that he would notice almost nothing except some deformations of his own body related to tidal effects (since the various parts of his body are subject to a slightly different attraction). But some

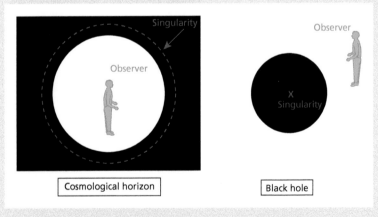

Figure VII.2 Comparison between the cosmological horizon and the horizon of a black hole.

researchers have argued recently that, when dealing with an event horizon (future horizon), he would come across a 'firewall' that would reduce him to dust. This led Stephen Hawking to suggest that the horizon surrounding black holes was an apparent horizon. An assertion that led the media to run as a headline: 'Stephen Hawking says black holes do not exist!' You see that the initial assertion was quite different. But the scientific debate remains open.

As far as I am concerned, I am convinced that there is much to learn by comparing the black hole horizon with the cosmological one. Their configurations are very different (Figure VII.2): in the cosmological case (respectively, the black hole one), the observer is at the centre (respectively, outside) of the region of space–time surrounded by the horizon. The singularity (the Big-Bang or the centre of the black hole) is on the other side of the horizon. There are some important similarities: in both cases, the question of the type of horizon (future or apparent) is raised; in both cases, we consider that Hawking radiation takes place at the horizon. Besides, drawing on the black hole case where all the information seems to be transcribed onto the horizon, Gerard 't Hooft and Leonard Susskind have suggested what is called the *holographic principle:* the whole information contained in our Universe would be encoded in our cosmological horizon! The latter would be a kind of hologram representing the whole Universe. The dynamics of this horizon would refer to the dynamics of our whole Universe.

This proposition, which seems astounding at first sight, has been unexpectedly supported by theoretical developments in string theory. This is what we call the duality between gravity theories in a space–time and gauge theories on the edge of this space–time (the horizon?). It goes a little beyond the scope of this book, but we can learn from it something important.

In the quest of the theory that will unify gravitation with the other fundamental forces, the gravitational horizon is very likely to play a central role. Is it not the place where Hawking radiation, an outstandingly gravitational quantum process, is emitted? The key to the fundamental questions that arise probably relies on a better comprehension of the dynamics of these horizons, but also on an experimental observation of their properties.

I can't finish without mentioning the final scene of a cult movie, *The Incredible Shrinking Man*, directed by Jack Arnold (1957), which perfectly summarizes my words. The main character, Scott Carey, after being exposed to an insecticide and then to a radioactive fog, begins to shrink. After falling into a cellar, he will have to fight a spider much bigger than he is. His tiny size will eventually allow him to escape by the grating of a basement window, which prevented him from doing so before. Being microscopic, he will be able to cross the horizon of the basement window and thus reach the infinitely large of our Universe, symbolized by the galaxies. What a beautiful metaphor of the two infinities we just discussed!

8

Gravity Turned into Waves

Like as the waves make towards the pebbl'd shore
WILLIAM SHAKESPEARE, *Sonnet 60*

Electromagnetic waves have allowed us to discover the breadth of phenomena associated with the electric and magnetic forces. They are produced by the motion of electric charges such as electrons in an antenna. Up to this day, they have allowed us to observe the Universe, first thanks to visible light, then through the entire electromagnetic spectrum. In a similar way, waves produced on the surface of a liquid by the fall of an object can give us information on the nature of this liquid, possibly even on its depth.

There are waves associated with gravitation, called gravitational waves. They are produced by the rapid motion of an important quantity of matter. Indeed, we have seen that matter curves space–time: matter in motion triggers a front (comparable to a wave) of curvature that is set in motion, just like the ripples produced on the surface of water by a falling stone. Gravitational waves are precisely the curvature waves. Because gravity is a very weak force, these waves can propagate over enormous distances (possibly the size of the observable Universe) without any deformation: matter encountered on the way perturbs them minimally. They are thus a remarkable means of observing all phenomena of gravitational origin, in particular black holes, which are the quintessential gravitational objects, and the whole Universe, which, as we have seen, is engineered by gravity.

The inherent difficulty with this type of observation is related to its exceptional scientific potential. I just said that the gravitational force is so weak that the associated waves propagate without being perturbed. For the same reason, they are very difficult to detect. This is why the direct detection of these waves had to wait 100 years and for the construction of incredibly precise detectors. The announcement on

11 February 2016 of the discovery of gravitational waves was a major event in the observation of the Universe, comparable to Galileo's telescope. For the first time, we directly observed the gravitational Universe and the phenomena that set this Universe in motion. It is this beautiful scientific adventure, climax of 100 years of instrumental, conceptual, and theoretical efforts, which I invite you to join in this chapter.

Waves of curvature

We have seen that gravitational waves are curvature waves. Because they curve space–time, they generate, wherever they go, a gravitational field that acts on all material bodies. Let us imagine that we have arranged into a circle of 1 metre diameter a collection of identical masses, as indicated in Figure 8.1, on the left side. When a gravitational wave is travelling orthogonally to the surface of the page, the distances between the masses evolve in such a way that the circle is deformed into an ellipse oriented alternatively along one direction, then along the perpendicular direction.

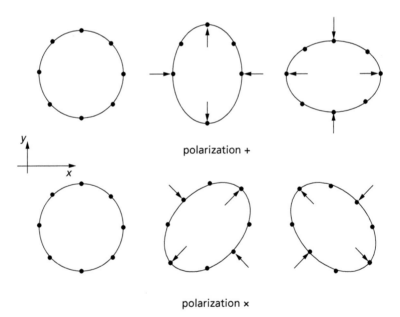

Figure 8.1 Apparent motion of a collection of identical masses with a passing gravitational wave of + (top) or × (bottom) polarization; the wave travels in the direction transverse to the page.

In general, the polarization of a wave is associated with the direction of the field that propagates. For example, for an electromagnetic wave such as light, there are two possible states of polarization related with the configuration of the electromagnetic field. A polarizer can extinguish one of the two polarizations.[1] Two different polarizers can completely stop a light ray. In the same way, a gravitational wave has two states of polarization, which correspond to the two deformations represented in Figure 8.1 (top and bottom, traditionally noted + and ×, respectively).

An immediate consequence of the motions presented in Figure 8.1 is that a solid body of non-negligible size is necessarily subject, when a gravitational wave is passing through, to forces which tend to deform it. These forces are in some ways similar to the tidal forces exerted by the Moon on Earth's oceans: they are called gravitational tidal forces. Figure 8.2 shows both the similarities and the differences between the two situations.

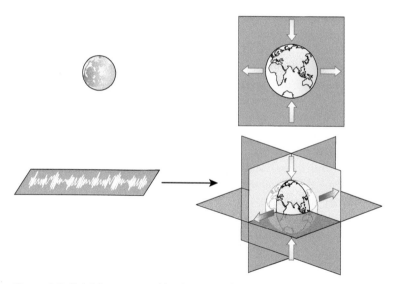

Figure 8.2 Tidal forces created by the Moon (top) and by a gravitational wave (bottom). The arrows indicate the direction of the deformation forces.

---

[1] An effect that is well known to photographers: reflection of light on glass or on a liquid surface selects one state of polarization. To decrease reflection, for example in order to take a picture through a window, the photographer uses a polarizer to stop the reflected light. The windshield of a car is also polarized to extinguish the reflections of light on a wet road.

In particular, whereas the lunar tidal forces are exerted in the three directions (by stretching in the Moon direction and by compressing in the two orthogonal directions), we have seen that the gravitational tidal forces are only exerted in the two directions transverse to propagation of the wave. Moreover, they are only active when the wave is passing.

A wave, whether it is light, sound, or radio, is characterized by its amplitude, its velocity, and its frequency or wavelength (if you know the velocity, then the frequency is obtained by dividing velocity by wavelength).

The *amplitude* of a wave measures the magnitude of the physical effect by which the wave is manifested. For the gravitational wave that passes through our page and through Figure 8.1, this is simply the relative variation of distance between the masses arranged in circle, i.e. the ratio between the variation of distance and the global size of the circle. The gravitational waves that are typically searched for have an amplitude on the order of 10^{-21} to 10^{-24}; in other words, if the circle has a diameter of 1 m, the masses move with respect to one another some 10^{-21} to 10^{-24} m; if the circle has a diameter of 1 km, the masses are moving from one another some 10^{-18} to 10^{-21} m. We could be surprised to see that the masses seem to move differently depending on the size of the circle although the same gravitational wave is passing through, but let us remember that there is no force acting, only the deformation of space–time which modifies the distances between two objects: a mass appears to move differently with respect to another mass depending on its distance to this mass. Such a behaviour shows that it is the texture of space–time that is deformed; the masses are only here to display the phenomenon.

These variations are incredibly small. This is due both to the weakness of the gravitational force and to the large distance of the sources of gravitational waves. Indeed, because a gravitational wave develops in all directions it loses intensity as it goes away from the source (its energy is spread on a larger and larger sphere) and its amplitude decreases as the inverse of the distance to the source. Thus, whereas the estimates above imply an amplitude of 10^{-14} m for the tidal oceanic motion associated with the passage of a gravitational wave (much smaller than the lunar tidal forces), the same gravitational tidal forces would be sufficiently important to break up Earth if it came close to the horizon of a supermassive black hole.

Doubts about the existence of these waves

As soon as the existence of gravitational waves was predicted by Einstein, some physicists considered that these waves were not real but simple mathematical artefacts, due to a wrong interpretation of the equations: a simple definition of space and time coordinates would make them disappear. Because these kinds of arguments have lasted, and because Einstein himself was caught up in them for a while, it might be interesting to say a few words about them.

The term 'gravific waves' appeared in a 1905 article by the French mathematician Henri Poincaré where he tries to include gravity into his own work on special relativity. But in June 1916 Einstein showed that the equations of general relativity imply the existence of gravity waves or gravitational waves. In the same article he computes the energy lost by a system in the form of gravitational waves. He first believed that a system with spherical symmetry can emit gravitational waves, before identifying his mistake and correcting it two years later, in what has been called since his quadrupole formula. Let us note in passing that Einstein's result, for example, imposes that a supernova must explode in an asymmetric, i.e. nonspherical, way in order to be a source of gravitational waves.

Did you say quadrupole?

As its name indicates, a quadrupole is a distribution of mass to four poles disposed in four quadrants. The fact that a source of gravitational waves must have such a structure at least (or even more complicated structure: octupole, etc.) is not surprising when you look at the structure of the wave produced (Figure 8.1).

For reasons that it would be good to clarify, maybe due to the conviction that gravity behaves differently from other interactions, a certain number of scientists have doubted the existence of these waves, waves that 'propagate at the speed of thought', said Arthur Eddington in 1922. In 1937, Einstein himself, in an article written with Nathan Rosen, briefly changed side before recognizing his mistake and correcting the article, much to the annoyance of Rosen. The debate, however, continued for a certain number of years, partly closed by a very simple argument proposed by Richard Feynman in 1957, and popularized by Hermann Bondi, the argument of the 'sticky bead'.

Mr Smith and the sticky bead

In 1957, a conference took place in Chapel Hill, North Carolina, on 'the role of gravitation in physics'. A topic less fashionable than today since Richard Feynman, the great American physicist, never short of a good joke, registered there under the name of Mr Smith. He proposed a thought experiment that made a deep impression on everyone. He considered a simple system of a bead freely sliding along a rod, although with a little friction. He presented this rod as being perpendicular to the direction of propagation of a hypothetical gravitational wave. As we have seen, the rod is subject to tidal forces that tend to deform it, by alternately contracting and stretching it. The small bead follows the motion set by the wave, but, since it is (almost) free, it slides along the rod and the friction that results produces some heat.

The wave has now passed. The system gets back to its original state but some heat has been produced. This is an unambiguous sign that the gravitational waves which produced it is not just an illusion. The wave exists and carries energy: it loses a very small fraction of it when going through the rod/bead setup.

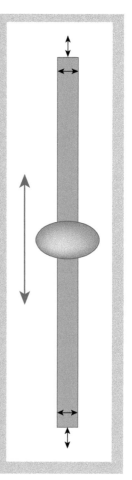

At the speed of light

To obtain the velocity of gravitational waves, we may need to come back to what we said in Chapter 6 on the link between the mediator of a force and the range of this force. We have seen that a force such as the gravitational force is of infinite range because the mediator, here the graviton, has vanishing mass. But relativity tells us that a massless particle propagates at the velocity of light.

Every electromagnetic phenomenon can be considered as resulting from the propagation of photons: similarly, every gravitational phenomenon, in particular a gravitational wave, can be interpreted as resulting from the propagation of gravitons. We deduce that gravitational waves travel at the speed of light at least in the framework of general relativity.

Does this mean that, in case of a cataclysmic event in the Universe, gravitational waves and electromagnetic waves would arrive at the same time? Not necessarily, because light or more generally electromagnetic waves can be retarded by the matter surrounding the event: imagine, for example, that the event is surrounded by dust clouds across which light must diffuse (i.e. be reflected from grain of dust to grain of dust in a zigzag trajectory); it will have to travel a distance longer than a straight line to reach us. In contrast, the gravitational wave superbly ignores these grains of dust and travels in a straight line. It reaches us first.

Let us now come to the wavelength or frequency of gravitational waves. In a way similar to electromagnetic waves that cover 20 orders of magnitude in frequency, from radio waves to gamma rays, the expected gravitational waves cover as well nearly 20 decades (Figure 8.3). Each region of wavelength is associated with a particular type of astrophysical source. We will return to them throughout this chapter.

An important example of an astrophysical source is provided by binary compact sources, which are sources composed of two compact stars, neutron stars, or black holes. We could imagine that binary astrophysical sources are rare. But this is not the case: one star out of two that we see in the sky has at least one companion star. For example, the closest star to the Sun, Alpha Centauri, is actually a system of three stars: another illustration of gravitation's importance in our Universe.

Two stars rotating around one another are just mass in motion. The binary system thus emits gravitational waves. We easily understand that the frequency of the emitted gravitational waves is twice the rotation frequency of the system, at least if the masses are identical: for a distant observer, the system returns to the same configuration at each half-period of rotation (Figure 8.4a and b).

The laws established by Johannes Kepler give this rotation frequency in terms of the common mass M and the size R of the binary system (the square of the frequency is proportional to the mass and inversely proportional to the cube of the distance, the proportionality coefficient being given by Newton's constant). This allows us to identify the frequency expected for the gravitational waves emitted by such binary

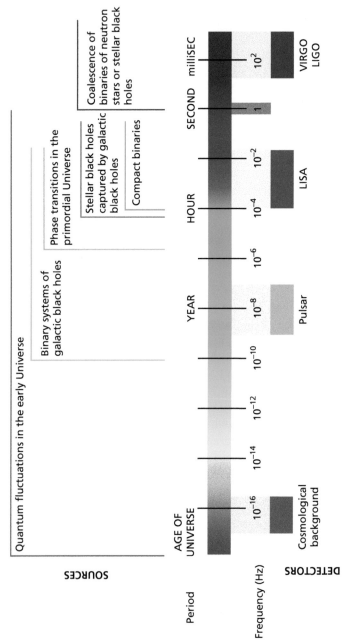

Figure 8.3 Frequency spectrum expected for gravitational waves produced in the Universe.

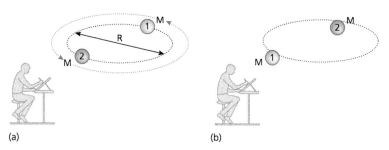

(a)   (b)

Figure 8.4 Binary system (identical masses) as seen by a distant observer.

systems. Let us take a binary of neutron stars (the mass M is typically 1.4 times the mass of the Sun and the distance (R) is on the order of 100 km), the frequency is on the order 100 Hz. We will see later that the large terrestrial detectors aim at this type of frequency. If, on the other hand, we take a binary of supermassive black holes, such as those that form in galaxy collisions (as we will see in the next chapters), then M is on the order of a few million solar masses and R on the order of an astronomical unit (the Earth–Sun distance, i.e. 1.5×10^{11} m): we are then in the range of 10^{-4} Hz, aimed at by the large space detectors.

For the sources of cosmological origin, it is the size of the cosmological horizon that fixes the wavelength of the gravitational wave. For example, this horizon size is 100 million light years ($1°$ in the sky, see Figure 5.2) at the time of recombination: this gives a frequency of 10^{-16} Hz. Events generating gravitational waves at even earlier times correspond to smaller horizons, thus smaller wavelengths and higher frequencies. We thus see that a very large spectrum of frequencies is anticipated for the frequencies of gravitational waves.

Cosmic Pas de Deux

Let us concentrate one moment on a system of two black holes that can be of stellar origin or supermassive black holes at the centre of galaxies. Because the system is losing energy in the form of gravitational waves, the two black holes get closer, and according to Kepler's laws, their rotation frequency increases. This process goes on until the black holes are so close that their horizons touch: the two black holes then form a single, they merge.

This fusion of two black holes is quite a remarkable event. Black holes being quintessential gravitational objects, their fusion represents a major opportunity to test the predictions of the theory of gravity under extreme conditions. For example, what happens when the horizons of two black holes make contact? What is the mass of the final black hole? More generally, since each black hole has three hairs according to the no-hair theorem (mass, charge, and angular momentum), this makes six hairs in total, hence three too many for the final black hole: how does it lose these extra hairs? And what is the fate of the matter that surrounds the two black holes?

It is thus absolutely remarkable, and a real scientific opportunity, that an event of this type was at the origin of the discovery of gravitational waves. I will come back to it in detail in the next chapter. For the time being, let us examine closer the different phases of this cosmic *pas de deux*.[2] They are summarized in a famous diagram thanks to the American physicist Kip Thorne (Figure 8.5).

We have already identified the first phase called spiralling. Because some energy is lost, the two stars get closer in a motion similar to a spiral and their rotation frequency increases: the frequency of the emitted gravitational waves increases accordingly. The exact form of the waves can be computed in an approximation of general relativity called post Newtonian—well adapted here because the two black holes, as long as they are not very close, can be likened to traditional stars.

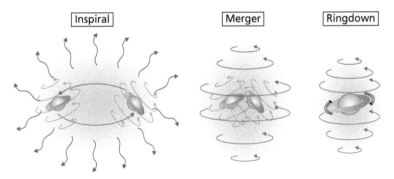

Figure 8.5 The three phases of the coalescence of two black holes according to Kip Thorne: on the left the spiralling phase, in the centre the plunge or merger, and on the right the ringdown phase.

---

[2] *pas de deux* is a duet sequence in a ballet.

At a certain moment, the two black holes become sufficiently close for their horizons to make contact. They can no longer be treated as simple massive objects due to the geometry of space–time very close to their horizons becoming nontrivial and the location of phenomena that we mentioned in the previous chapter. Because the gravitational field is intense in these regions, the approximation methods favoured by physicists are no longer valid. We then need to treat the problem in a numerical way, i.e. on a computer.

The task turns out to be extremely difficult: Einstein equations actually represent 10 coupled equations, depending on four space–time variables (one for time and three for space). In 1995, American teams set up the Binary Black Hole Grand Challenge Alliance to take up the challenge of black hole binaries, by pooling unprecedented computing resources. But no significant results were obtained.

However, 10 years later, in 2005, a researcher from Princeton University, Franz Pretorius, completed the tour de force of providing the first numerical simulation of the complete merger of two black holes, with an evaluation of the gravitational waves produced. And the situation must have been ripe since this premiere was quickly reproduced by several teams in the USA and in Europe.

An example of such a simulation, although more recent, is shown in Figure 8.6. We see how space–time is carried along into the spiralling motion that accompanies the final plunge of the two black holes towards one another. The fine understanding of this phase made it possible to deduce precise predictions on the form of gravitational waves, which are then produced. This is very precious information because it gives us first-hand knowledge about a regime of gravitational interaction that we know poorly, and on the form of space–time under these extreme conditions.

The last phase, called ringdown, is due to the fact that the astrophysical object obtained during the coalescence is still too complex to be a simple black hole. We have seen in particular that it has too many 'hairs', i.e. too many characteristics. How are they lost? In the form of gravitational waves. It is thus a phase of simplification where the black hole loses its last characteristics. Imagine, for example, two bubbles of air in a liquid that we make coalesce: the object thus formed will be a place of transitory oscillations before turning into a perfectly spherical bubble. These transitory oscillations will dissipate energy into the surrounding liquid just like a merged black hole does in space–time.

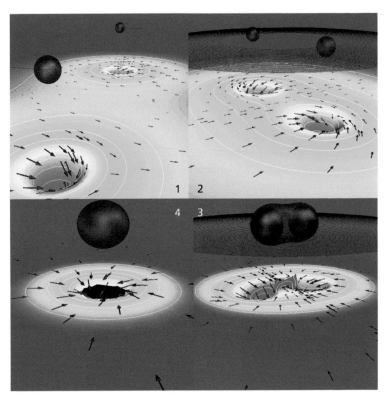

Figure 8.6 Successive phases (clockwise) of the final plunge of two black holes obtained by a numerical simulation of the SXS (Simulating eXtreme Spacetimes) collaboration. The horizons are represented by black spheres, below which the corresponding space–times are sketched.
© Caltech-Cornell.

In total the gravitational wave produced has the form given in Figure 8.7. The characteristic increase of the wave frequency at the time of the plunge led physicists to compare it to the chirp of birds, the short increasingly high-pitched noise they make. Indeed, if we transcribe the signal into sound waves, each coalescence of black holes sounds like the chirp of a robin! The parameter that describes this increase is a combination of the masses of the two black holes, called chirp mass.

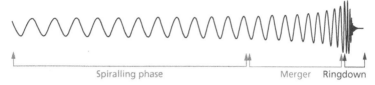

Spiralling phase Merger Ringdown

Figure 8.7 Form of a gravitational wave produced during the final phases of the coalescence of two black holes.

The study of how the frequency varies with time makes it possible to identify the parameters of the source, in particular the mass of the initial black holes and of the final one. This then makes it possible to determine the magnitude of the gravitational signal at the source. Indeed, as we have seen, because it develops in all directions, a gravitational wave loses intensity when moving away from the source: its amplitude decreases inversely proportional to the distance from the source.

I have until now considered ideal black holes which appear in general relativity. But we have seen that astrophysical black holes are surrounded with matter: accretion disks, jets of particles, etc. We could wonder where this matter is going to perturb the signals that we have identified. It is not the case, because this matter remains at distances from the horizon much larger than those we consider in the last stages of the coalescence of two black holes. On the other hand, the energy produced in the form of gravitational waves is enormous (much larger, for example, than all the luminous energy produced by the stars of our galaxy). If a very small fraction of its energy is converted into luminous (or electromagnetic) energy when it travels through this zone of matter, then this will be a signal complementary to the gravitational signal, less direct but precious nevertheless because it can give us information on the physical processes at play around black holes. The difficulty is that the electromagnetic signal can take years to emerge from matter and thus is not necessarily simultaneous with the gravitational signal (gravitational waves move through the layers of matter without interacting significantly and thus are not retarded).

Before turning to the discovery of gravitational waves, let us see the means of the discovery: these gravitational wave antennas, outcome of 100 years of research (for an historical perspective on these efforts, see Focus IX) that overcame incredibly difficult technical challenges!

How to identify gravitational waves?

I have emphasized several times the ambivalence of the detection of gravitational waves: because the gravitational interaction is very weak, these waves propagate basically without any deformation over distances on the order of the size of the observable Universe; for the same reason, their effect on the matter that is around us is infinitesimal. This can be quantified: I have mentioned variations of distance from 10^{-18} to 10^{-21} m for objects located at a distance of 1 km!

These orders of magnitude are extraordinarily weak. The task could even seem impossible when we think about the size of the proton, about 10^{-15} m. But we must remember that it is the very large collection of atoms of a material body (about 10^{24} for a few grams of matter) that is displaced by the same distance with respect to another body!

Nevertheless, methods of ultraprecise metrology must be implemented in order to achieve such precision. We appeal to the best length standard that we know: light. We have seen in Focus II that its velocity makes it possible to define the metre. And it is the period[3] of the light emitted by an atom of cesium 133 (in a specific atomic transition) that makes it possible to define the second. How do we implement such a very precise standard as a light wave? By using a technique based on the phenomenon of interference.

Precision metrology and interferometry

Let us see more closely what is the phenomenon of interference between two waves. In order to do so it is more intuitive to first consider the waves at the surface of a liquid. Imagine that you throw simultaneously two stones into a quiet pond. Starting from the impact of each stone, concentric waves form on the surface of water, the wavelength of which is directly related to the size of the stone. In the zone between the two impacts, the waves produced by each of the stones overlap and you can see the development on the surface of water of a pattern characteristic of what one calls the phenomenon of interference (Figure 8.8). If you throw many stones in an incoherent way, the figure of interference is blurred and is replaced by choppy waves on the pond.

---

[3] The period is the time it takes light to travel a distance equal to the wavelength.

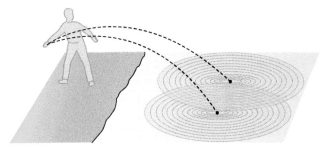

Figure 8.8 Two stones thrown simultaneously into a quiet pond produce waves on the surface that merge into an interference pattern.

The same phenomenon can be reproduced with light waves as in the double-slit experiment reproduced in Figure 8.9. A single light wave is projected onto a plate pierced with two slits; the two waves that come out on the other side overlap and produce a figure of interference on a screen placed further away; a succession of dark and bright bands,

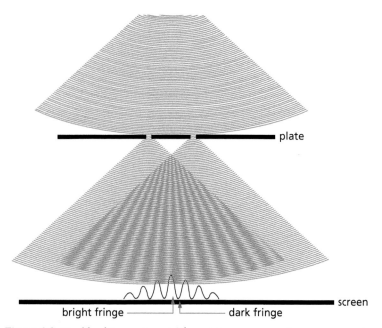

Figure 8.9 Double-slit experiment. The minima, resp. maxima, correspond to regions on the screen where the waves interfere destructively (dark fringes), resp. constructively (bright fringes).

called fringes, are observed on the screen. It is important that the light source has the property of coherence mentioned earlier, a property that is found, for example, in laser light.

The interference pattern depends on the distance scales that are characteristic of the experiment. If we move the screen further away, the fringes get further apart. Conversely, if we move the two slits further apart, the fringes get closer.

This principle makes it possible to measure very weak variations of distance. It was applied in the setup called the Michelson interferometer, which allowed Michelson and Morley to realize their historical experiment on the velocity of light. The principle is summarized in Figure 8.10: a laser beam falls on a semi-reflective mirror called a beam splitter. Each split beam travels a similar distance before being reflected by a mirror to return to the beam splitter; the two beams are then recombined and the corresponding light waves interfere. The interference pattern is observed on a screen.

If the travel distances in the two arms of the interferometer are exactly equal, the two waves are identical when recombining: we have a bright spot at the centre of the image on the screen. As we move away from this centre, the interference is alternatively destructive or constructive (Figure 8.9), which we correspondingly encounter in successive dark and bright fringes. If the travel distances differ, which we can obtain by acting delicately on the mobile mirror, then we see a certain

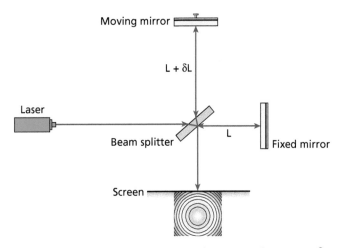

Figure 8.10 Principle of the Michelson interferometer. The quantity δL is the difference between the distances covered by the beams in the two arms.

number of fringes scrolling past, a number equal to the ratio of the difference of arm lengths to the light wavelength.

Thus, when we use a laser with a wavelength of 1.064×10^{-6} m, if we see two fringes scroll by, we can conclude that the mirror moved by 2.128×10^{-6} m.

We are still far from the 10^{-18} m aimed at, but this shows that Michelson interferometers are a sound base for achieving precise metrology. Before describing their use for gravitational waves, let us return to the specifications needed for such detectors.

Which size for the detectors?

In order to detect gravitational waves, we must be able to measure the motion of two masses placed at a distance on the order of the associated wavelength (or at least not much smaller).

This wavelength is obtained from the frequency of the wave and its velocity: more precisely it is computed by taking the ratio of the velocity of light (in km/s) to the frequency (in Hz). We saw in the preceding chapter that in the case of a binary system of neutron stars the frequency of the gravitational wave produced is on the order of 100 Hz. The wavelength is thus of the order of 300,000/100, i.e. 3,000 km, a size rather huge for an interferometer: remember that the one used by Michelson and Morley fit on a table!

For a binary system of supermassive black holes, the frequency is on the order of 10^{-4} to 10^{-2} Hz and thus the wavelength is greater than 30 million km. Needless to search for a site on Earth to build a detector whose size would not be ridiculously small compared to such a scale. But the fearless hunters of gravitational waves did not stop at this minor detail: space is big enough to accommodate such a monster!

The advantage of going into space is also to free oneself from seismic waves that have the unpleasant property of being in the same frequency band: it would be unfortunate to mistake one of these seismic waves for a gravitational wave.

Interferometers ready to detect gravitational waves

The idea to use interferometry to detect gravitational waves goes back to the end of the 1950s. However, the first real design dates from the beginning of the 1970s, on the impulse of Rainer Weiss at MIT. In parallel, Ronald Drever in Glasgow conceived the idea of adding a resonant cavity to enhance the performance of the interferometer. He joined Caltech in 1979, where the theorist Kip Thorne was already, to build a prototype that would give rise to the Laser Interferometer Gravitational-Wave Observatory (LIGO) interferometer. In 1983, Alain Brillet developed, in Orsay (France), an interferometer model that would become the French-Italian VIRGO detector. Currently, terrestrial gravitational antennas based on this concept have been developed all around the world: besides the two LIGO interferometers in the United States, and VIRGO in Italy near Pisa, we also count GEO600 near Hanover, Germany, KAGRA in Japan, and soon a LIGO interferometer in India. What motivated such a worldwide interest, when gravitational waves were still waiting for a direct detection?

Before answering this question, let us look a little closer at what an interferometer for the detection of gravitational waves looks like. It is

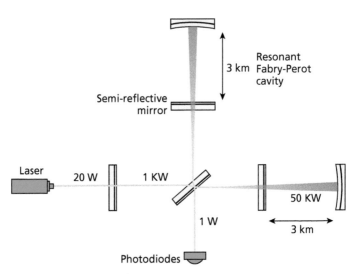

Figure 8.11 The interferometer used by the VIRGO experiment. We can see on each arm the resonant cavity (called Fabry–Perot). A recycling mirror, placed on the incoming beam, allows the power inside the interferometer to increase by a factor of 50. Interference fringes are measured by photodiodes (at the bottom).

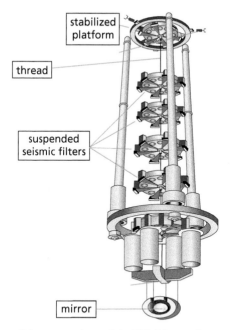

stabilized platform

thread

suspended seismic filters

mirror

Figure 8.12 One of the suspensions of the VIRGO interferometer that allows for strong absorption of seismic noise. © EGO/VIRGO.

actually very similar in its principle to the Michelson interferometer (Figure 8.10). The main addition (Figure 8.11) consists of resonant cavities, which are made with semi-reflective mirrors on each arm: light finds itself trapped between the mirrors at the end of each arm and the semi-reflective mirrors. Light goes back and forth many times inside the 'cavities' thus formed (on average it stays there for about one thousandth of second and thus travels 300 km). This clever trick, originally due to Drever, makes it possible to lengthen the effective distance travelled by light. It is also used to stabilize the laser light.

We have seen that a Michelson interferometer makes it possible to measure distances on the order of 10^{-6} m, and we must still gain some 12 orders of magnitude. The developments mentioned and many others make it possible to successfully meet this challenge.

If a gravitational wave passes through such an interferometer, for example orthogonally to this page, then all its components move as in Figure 8.1. This means that the length of one arm increases when the other one decreases, which makes the interference fringes scroll in a periodic way in the light detector (photodiode). It is in this way that we are able to detect a passing gravitational wave.

Of course, at such a level of precision we must fight against numerous sources that come to perturb the measurement (called noise). The most difficult to control, especially at low frequency, are those arising from seismic wave types (earthquakes or simply a truck or train passing by). This is, for example, why the elements of the detector hang from impressive suspensions that isolate them from the major part of seismic noise (Figure 8.12).

The astrophysical events that these detectors aim to identify are, in particular, asymmetric explosions associated with supernovae or gamma ray bursts, or the fusion of two neutron stars, two black holes, or one neutron star and one black hole. In all cases, these astrophysical objects must typically have the mass of a star (such as the Sun) for the event to be observable in the detector frequency domain.

It is important to identify the position of the sources in the sky, in particular to see whether they also emit electromagnetic radiation, or high-energy particles (observed by detection on Earth). For this purpose, a method that follows the triangulation method invented by Thales 2600 years ago is used. In order to measure the distance to a boat visible from the shore, Thales placed two observers A and B at a given distance from one another and asked them to note the angle between the line between them and the direction along which each of them sees the boat (Figure 8.13). By reconstructing the triangle with one side (AB) and two angles, he deduced the distance from the boat to the coast.

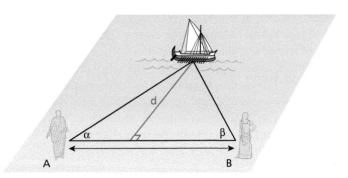

Figure 8.13 Triangulation according to Thales. The position of the boat, from the shore, is determined by the measurement of the distance AB and the two angles.

In a similar way, we can use two (or more) interferometers located in two different places in order to triangulate a cosmic source and obtain a precise estimate of its position. This is why the LIGO collaboration built two detectors, one in Livingston, Louisiana, the other 3,000 km away in Hanford, Washington. But in order to have a better precision on the direction, it is important to have a reference distance as large as possible, and thus to have gravitational antennas scattered all over Earth: this is why the LIGO collaboration has offered to build one of its detectors in India.

We have seen that the frequency domain of terrestrial detectors, from 10 to 1,000 Hz, corresponds to sources where the displaced masses are on the order of the mass of the Sun. This gives fairly weak amplitudes at the origin of the event, which are detectable only at a distance fairly close to the source (we have seen that the amplitude of the signal decreases with distance). This is why the first versions of the LIGO and VIRGO detectors were only sensitive to vicinity sources (on the order of 20 Mpc for a binary system of neutron stars). An important effort has been made to increase the sensitivity of these interferometers, which are now called Advanced LIGO and Advanced VIRGO (Figure 8.14), to

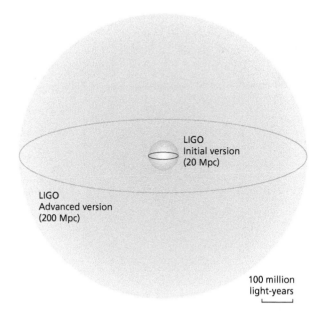

Figure 8.14 Detection zone for binary system of neutron stars of the LIGO detector in its initial (20 Mpc) and in its advanced (200 Mpc) versions.

some 200 Mpc. By multiplying the distance by a factor of 10, we have multiplied the accessible volume by 1,000.

To predict the number of events expected for a given detector, we must have an estimate of the event rate, i.e. the number of events per year and for a certain volume. Astrophysical models made it possible to estimate these rates but with important uncertainties: let us say that the rates are estimated to within a factor 10 to 100. The first generation of detectors, even if it was obtained by a technological tour de force, reached a sensitivity which ensured a detection only in the case of an optimistic prediction for these rates.

The new generation, called advanced, gave good hope that, within a few years, detection would be made: thanks to improvements in sensitivity, the multiplication by 1,000 of the observed volume led to the hope that detection was possible even in the case of a pessimistic prediction for event rates. Nature proved to be generous for physicists …

Imprints of the primordial gravitational waves in the Cosmic Microwave Background (CMB)

One of the most violent events in the Universe that comes to mind is the phase of inflation, which started the Universe's evolution just after the Big Bang. Gravitational waves were produced continuously during inflation, and also at the end of inflation—a phase called reheating.

I have described gravitational waves as waves of the space–time curvature. Yet, we saw in Chapter 5 that inflation tends to flatten any curvature. This seems to be contradictory. However, recall that quantum fluctuations tend to perturb very slightly the geometry of space–time. Indeed, they precisely reintroduce curvature deformations that propagate as gravitational waves. We thus expect, in addition to the observed perturbations of CMB radiation, the presence of gravitational waves produced during this phase. The detection of primordial gravitational waves is crucial because only their discovery would confirm that the observed CMB fluctuations are indeed related to gravitation and to the structure of space–time, more specifically to the haziness of this structure introduced by quantum physics.

One way to detect these primordial gravitational waves emerging from inflation is to use their interaction with CMB photons. This interaction generates a polarization of CMB light! To understand this, let us recall that photons are sensitive to the gravitational field (this is why the propagation of light is curved by the presence of a large mass), which means that photons are sensitive to the space–time deformations that gravitational waves propagate. Let us also recall that gravitational waves exist in two states of polarization (Figure 8.1), similarly to the photons that constitute light. It turns out that the interaction between gravitational waves and photons is such that the polarization of the former can influence the polarization of the latter: primordial gravitational waves thus polarize the cosmic background radiation.

This is why the ultimate goal of all experiments, both present and future, that scrutinize the cosmic background is to identify its polarization. This would represent a remarkable milestone, as it would mean breaking through the opaque wall that hides what happened before the epoch of recombination: gravitational waves produced some 10^{-38} second after the Big Bang would thus be detected.

But the task is very difficult. First, because we observe the photons of the cosmic background only today, which means only after they have been travelling for some 15 billion years: the influence of matter, such as galaxies, encountered along the way has modified their polarization. The installation of huge cosmological surveys, which are more and more precise and explore deeper and deeper (that is, see farther and farther), enables us to precisely estimate the effect of matter on polarization. But polarized photons can also be emitted by galactic dust.

This explains why the announcement, made in early 2014, by the American collaboration BICEP2 (Figure VIII.1) of the discovery of the polarization of the cosmic background by primordial gravitational waves was received with caution by the scientific community. The experiment's data came from a very small region in the sky. To be able to distinguish between the polarization arising from dust and that from astrophysical sources, it is crucial to observe the whole sky and in several wavelengths (because the dependence of other sources of polarization on wavelength differs from that of primordial gravitational waves). This is exactly what the Planck mission did and this is why the BICEP2 data have eventually been combined with the polarization data from Planck, with the conclusion that no significant deviation due to gravitational waves has been detected.

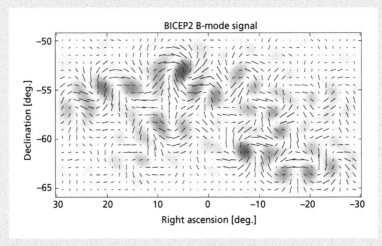

Figure VIII.1 Representation of the CMB light polarization, as observed by BICEP2 Collaboration in a small part of the sky. © BICEP2 Collaboration.

Despite this inconclusive result, it is nevertheless clear that the large media coverage of this BICEP2 announcement is a telling sign of the deep interest that a detection of primordial gravitational waves would generate.

What fraction of this polarization is due to intergalactic dust? And what fraction is due to gravitational waves that originate from the fluctuations of curvature in the primordial Universe?

9

We Did It!

A screaming comes across the sky. It has happened before, but there is nothing to compare it to now.

THOMAS PYNCHON, *Gravity's Rainbow* (1973)

On 11 February 2016 the room of the National Science Foundation (NSF) in Washington, DC, was full. It had been a few months since the rumour had gone viral on the web: the LIGO detector had barely started again taking in scientific data in its advanced version when it had detected an event proving the existence of gravitational waves. Astrophysicists in need of publicity, not necessarily the best informed, had published on blogs what they thought they had learned, but the LIGO collaboration itself had remained silent.

First, there was a brief introduction by Frances Cordoba, Director of NSF, reminding everyone what is at stake with gravitational wave research, then David Reitze, the LIGO director, stepped forward: 'We ... have detected ... gravitational waves. We did it!' A round of applause and flashes from photographers eager to catch this moment filled with emotion. Then Gaby Gonzalez, spokesperson for the LIGO Scientific Collaboration, Rainer Weiss, and Kip Thorne, whose pioneering role we have already mentioned, provided details about the discovery. And this discovery was as spectacular as the stakes were high.

1.3 Billion years ago

Two black holes whose respective masses are about 30 times the mass of the Sun have been rotating around one another since time immemorial. The binary system that they form has continually lost energy into gravitational waves. Thus they came closer and closer and rotated faster and faster. The moment that is of interest to us is a fraction of a second before their two horizons touched. The gravitational field between them was then very intense. Imagine about 60 solar masses localized in a region of a few hundred kilometres! Space and time were distorted by this concentration of mass and this distortion was propagating in all directions, in the form of gravitational waves of large amplitude. Everything happened very quickly. The two black holes merge into a single one, whose horizon keeps in a first stage the memory of its dual origin. But very soon, just as when two drops of mercury coalesce, the horizon undergoes oscillations, loses its last characteristics in the form of gravitational waves, and recovers its symmetric shape. The brand new black hole still curves space–time but no longer emits gravitational waves.

As much as 1.3 billion years later, on 14 September 2015 at 9 hour 50 minutes 45 seconds UTC exactly this distortion reached Earth. It was first detected by the LIGO Livingston detector in Louisiana, then 17 ms later by the Hanford detector in the state of Washington 3,000 km away. In the two cases, the signal stood very clearly apart from background noise. In fact, the LIGO physicists did not expect the existence of such massive binary sources: the signal thus had an amplitude much greater than expected. The automatic system of data processing started immediately and identified the signal's form. This signal was identical in the two detectors located 3,000 km apart. On the evening of September 14th, LIGO physicists were convinced they had made a major discovery.

You can see in Figure 9.1 this magnificent signal, as shown during the press conference of February 11 and in the article published in *Physical Review Letters* by the LIGO and VIRGO[1] collaborations and its comparison with theoretical predictions. In the upper panels, you can see the signals such as they were detected in Livingston (right) then in Hanford (left): the Hanford signal has been reproduced on the right as well; see how the two signals overlap! In the bottom panels, you can see in grey the signal that has been extracted by diverse data analysis methods (in particular a classical 'wavelet' method). You can recognize the classical chirp signal (frequency increase) that we encountered in Figure 8.7. This allowed the LIGO physicists to quickly deduce the chirp mass and thus to obtain an order of magnitude of the mass of the two initial black holes, some 30 solar masses. This was a surprise for many who considered that, for such a stellar system, the mass would not go

---

[1] Thanks to an earlier agreement between the two collaborations, the VIRGO physicists are analysing the LIGO data as well.

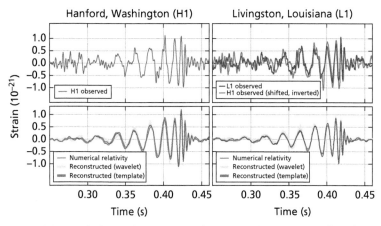

Figure 9.1 Signal observed on 14 September 2015 in the Livingston (right) and Hanford (left) detectors. On top, the signal observed; in the bottom, the signal reconstructed by two different methods, compared with the physical predictions (obtained by numerical relativity). © B. P. Abbott et al. LIGO Scientific Collaboration and Virgo Collaboration.

beyond about 10 solar masses, but an excellent surprise since it meant that the event was more energetic than expected.

A more systematic study allowed a more precise determination of the individual masses of the two black holes, respectively 29 and 36 solar masses, whereas the final black hole has 62 solar masses. Added up, 3 solar masses are missing. This corresponds to emission in the form of gravitational waves of some 10^{50} watts, in a few tenths of a second: more than what all the stars in the observable Universe emit during the same time in luminous energy!

Once these parameters were known, the signal's amplitude at the source could be determined. The amplitude of the signal observed then made it possible to deduce the distance from the source, 1.3 billion light years or 410 Mpc. This is larger than the 200 Mpc announced earlier (see Figure 8.14), simply because the binary system is much more massive (two black holes of some 30 solar masses instead of two neutron stars of 1.4 solar mass), hence the more energetic source.

Note the horizontal axis in Figure 9.1: everything happened in less than half a second. Isn't it incredibly short for a cosmic event? But you must remember that the LIGO detector is sensitive to waves in a certain frequency range. The frequency of the gravitational wave is directly related to the rotation frequency of the two black holes. Since, over millions of years, they have been moving closer towards one another, the frequency has increased until the wave is perceived by a detector (when this frequency reaches 10 Hz), which occurs for only a fraction of second before the final plunge, and only if the event is powerful enough to be detected.

Is this an incredible fluke? Not really we may expect. Such events take place in many different locations in the Universe, and the detector only caught the gravitational wave that was passing on September 14.

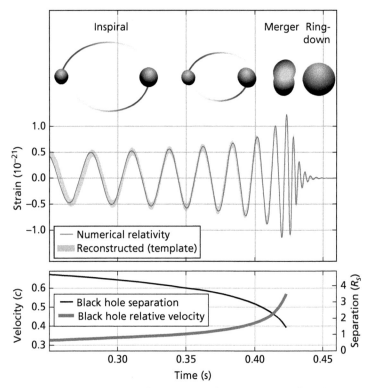

Figure 9.2 The three phases of the merger of two black holes (spiralling on the left, proper merger on the right, and ringdown far to the right). In the bottom panel, evolution with respect to time of the distance separating the two black holes (in black), in Schwarzschild radius units, and of the velocity of rotation (in green), in units of c, the speed of light. © B. P. Abbott et al. LIGO Scientific Collaboration and Virgo Collaboration.

We may expect to detect other events in other points of the Universe in the future.

We have seen in the preceding chapter the huge work carried out by theorists to predict the precise form of the signal for this type of event. In this precise case, you can see this in Figure 9.2 that comes from the discovery article.

You can see here the three phases and to which oscillations of the detector they correspond. In the bottom panel, you can see evolution with respect to time of the distance separating the black holes in multiples of the Schwarzschild radius of the final black hole, around 185 km. You can also see their relative velocity in units of the velocity of light: it evolves from 1/3 of the velocity of light to 2/3 of it at the time of impact! A really incredible event!

GW150914

This code name may not mean much to you but you may be certain that any physicist specialized in gravitational waves will become emotional at the mention of this series of letters and numbers. GW150914 is the name of a source observed by the LIGO detector in 2015, on September (09) 14. Obviously GW stands for gravitational waves. This was the first detection with gravitational wave, the first of a long series, let us hope. And the name of the following ones will follow the same pattern: GW followed by the date of detection.

If we take into account the fact that this event was produced as soon as the Advanced LIGO detector started taking data, and that the announcement on February 11th was only taking into account the analysis of the first 18 days of data, we can expect to discover many other sources during the coming years. Thus, we are witnessing the birth of gravitational astronomy. This will allow us to test general relativity in an increasingly precise way. On that subject, we should note that for the first time, the event GW 150914 makes it possible to test this theory in a regime where the gravitational field is intense. I have often insisted how the gravitational forces that we can experience are weak. But nothing like when two black holes of 30 solar masses move to a distance within a few hundred kilometres of each other. This is indeed why rotation velocities close to the velocity of light are reached! And, even under these extreme conditions, it appears that the theory of Einstein is verified.

It is the moment to pause and marvel at the fact that, at the beginning of the twentieth century, a physicist used his own very limited experience of the gravitational force, as well as a certain number of thought experiments, to conceive a theory that is verified today under extreme conditions that he could never have imagined (black holes were actually never his cup of tea).

But of course, we will need to check in a much more precise way that Einstein's theory is verified. We have seen that quantum phenomena such as Hawking radiation are taking place at the level of the horizon. We still ignore the quantum theory of gravity. Will the horizon behave as Einstein's classical theory predicts, or will fluctuations leave observable traces? Observation will bring answers to these questions that, not so long ago, were expected to remain for a long time as pure theoretical speculations.

2017: A Nobel Prize, colliding neutron stars … and this is only the start of gravitational wave research!

As of August 2017, three more gravitational waves produced by the coalescence of two black holes were detected. The last event on 17 August 2017 observed by LIGO and VIRGO made it possible to deliver a much better sky localization of the source.

With GW170817, for the first time, scientists directly detected gravitational waves in addition to light emitted from the collision of two neutron stars. Among the main results, this detection confirmed that the short gamma ray bursts are produced by the coalescence of neutron stars and that the gravitational waves travel at the speed of light.

On 4 October 2017, the Nobel Prize in Physics 2017 was attributed 'for decisive contributions to the LIGO detector and the observation of gravitational waves' to Rainer Weiss, Barry C. Barish, and Kip S. Thorne.

A hundred-year-long quest for gravitational waves

The theory of gravitational waves, as a consequence of general relativity, was conceived in two seminal articles by Albert Einstein: one in 1916, 'Approximate Integration of the Equations of the Gravitational Field' and especially one in 1918 'On Gravitational Waves' (Figure IX.1). The latter corrected a few mistakes in the first article and proved the famous 'quadrupole formula' that turns the energy loss of a binary system into gravitational waves. But few people, including Einstein himself, could imagine at the time that these gravitational waves would one day be discovered, so tiny was the estimated effect.

It was in the 1950s, in particular at the instigation of John Archibald Wheeler, that the experimental search for gravitational waves was really initiated. In 1955, the physicist Joseph Weber, from the University of Maryland, took advantage of a sabbatical leave to look into the question. During the next years he designed a detector that consisted of an aluminium cylinder, a kind of large bar. This detector started resonating for a specific oscillation frequency: if a gravitational wave with exactly this frequency were to pass through the detector, it would induce periodic deformations of the bar that then resonated; the effect would thus be amplified and registered by small sensors placed on the bar (Figure IX.2).

In 1969 Joseph Weber announced that he had discovered gravitational waves. The signal appeared to have been observed simultaneously in two bar detectors: one on the campus of the University of Maryland, the other one at the Argonne National Laboratory near Chicago. Unfortunately, nobody was able to reproduce later these results. Today, it is thought that they were due to a poor understanding of the instrumental noise of the detector.

Soon after emerged a new technique based on interferometry, first imagined by two Soviet scientists, Mikhail Gertsenshtein and Vladislav Pustovoit, then independently by the US and European teams, as we have seen in this chapter.

But, even if detection of gravitational waves was a long time coming, indirect observational proofs of their existence were obtained during

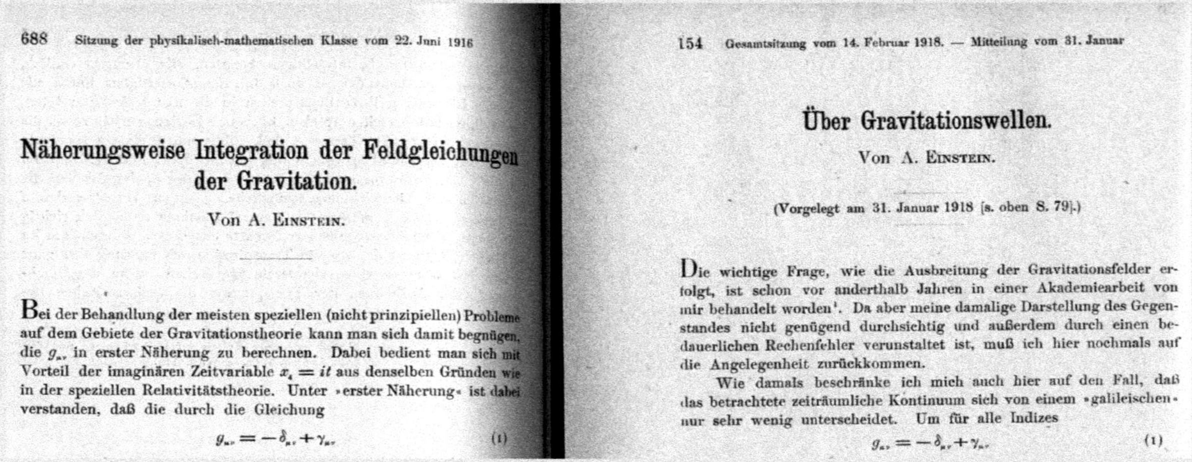

154   Gesamtsitzung vom 14. Februar 1918. — Mitteilung vom 31. Januar

Über Gravitationswellen.

Von A. Einstein.

(Vorgelegt am 31. Januar 1918 [s. oben S. 79].)

Die wichtige Frage, wie die Ausbreitung der Gravitationsfelder erfolgt, ist schon vor anderthalb Jahren in einer Akademiearbeit von mir behandelt worden. Da aber meine damalige Darstellung des Gegenstandes nicht genügend durchsichtig und außerdem durch einen bedauerlichen Rechenfehler verunstaltet ist, muß ich hier nochmals auf die Angelegenheit zurückkommen.

Wie damals beschränke ich mich auch hier auf den Fall, daß das betrachtete zeiträumliche Kontinuum sich von einem »galileischen« nur sehr wenig unterscheidet. Um für alle Indizes

$$g_{\mu\nu} = -\delta_{\mu\nu} + \gamma_{\mu\nu} \qquad (1)$$

---

688   Sitzung der physikalisch-mathematischen Klasse vom 22. Juni 1916

Näherungsweise Integration der Feldgleichungen der Gravitation.

Von A. Einstein.

Bei der Behandlung der meisten speziellen (nicht prinzipiellen) Probleme auf dem Gebiete der Gravitationstheorie kann man sich damit begnügen, die $g_{\mu\nu}$ in erster Näherung zu berechnen. Dabei bedient man sich mit Vorteil der imaginären Zeitvariable $x_4 = it$ aus denselben Gründen wie in der speziellen Relativitätstheorie. Unter »erster Näherung« ist dabei verstanden, daß die durch die Gleichung

$$g_{\mu\nu} = -\delta_{\mu\nu} + \gamma_{\mu\nu} \qquad (1)$$

Figure IX.1 Front pages of the two historical articles of Albert Einstein on gravitational waves (1916 and 1918).

Figure IX.2 Joseph Weber working on his detector bar in 1969 at the University of Maryland. © University of Maryland Libraries.

1970–80. The most convincing proof came from the study of millisecond pulsars: pulsars are neutron stars that spin very rapidly (in one thousandth of a second for a millisecond pulsar) and emit strong electromagnetic radiation in the direction of their magnetic axis. Because the magnetic axis does not correspond to the rotation axis, the electromagnetic beam sweeps the space around in a very regular manner, just like light from a lighthouse: an observer placed on the beam's trajectory thus sees it at very regular intervals.

Millisecond pulsars orbit around a companion most of the time. As we have seen, such a binary system in rotation emits gravitational waves; it loses energy and the two stars get closer; their frequency increases. This frequency drift can be measured very precisely in some cases and compared to the prediction of general relativity (which makes it possible to compute the loss of energy in the form of gravitational waves).

A Nobel Prize awarded for observational evidence of the existence of gravitational waves

The first study of this type was conducted by Russell Hulse and Joseph Taylor on the binary system PSR 1913+16 (PSR for Pulsar). It is a system of two compact stars of mass about 1.4 solar masses rotating around each other in an elliptic orbit (Figure IX.3). The system is very compact since the minimal distance between the two stars is on the order of the solar radius (700,000 km, i.e. 4 light seconds).

One of the two stars, the only one detected, is a pulsar, a neutron star emitting in our direction pulses in the radiofrequency domain every 59 thousandth of a second. We know that these two objects rotate around each other because fluctuations of this period of 59 thousandths of a second, depending on whether the pulsar is in a phase where it is moving away from us or in the phase where it is moving towards us, have been observed. Astrophysicists have deduced the binary system's rotation period to be 7 hours 45 minutes and 7 seconds, i.e. 27,907 seconds. This gives us the frequency of the gravitational waves that should be emitted by the system: 4.5030×10^{-4} Hz.

General relativity allows us to compute the energy lost by the system in the form of gravitational waves, and thus the drift of the rotation

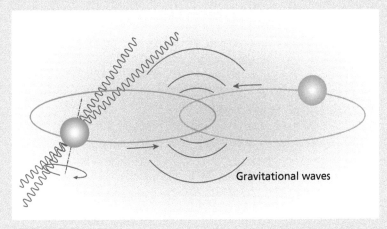

Gravitational waves

Figure IX.3 The structure of the binary system PSR 1913+16. On the left, the pulsar observed from Earth; on the right, its companion that remains unobserved from Earth (maybe also a pulsar, but emitting in a direction away from Earth).

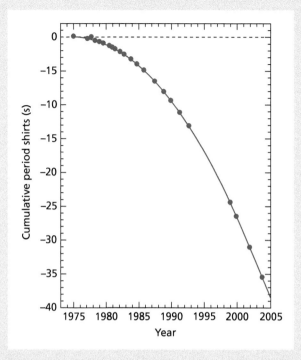

Figure IX.4 Cumulated decrease of the rotation period of the binary system PSR 1913+16 during the period 1975–2005. The dots correspond to the observational data, whereas the parabolic curve is the theoretical prediction.

period with time. According to detailed computations performed by the French theorists Thibault Damour and Nathalie Deruelle, it must decrease by 0.0765 thousandths of a second every year. This was verified by J. Taylor and J. Weisberg in a very precise way (Figure IX.4), ... which led to awarding the Nobel Prize to Hulse and Taylor in 1993!

10

A New Constellation in the Sky: LISA

If we have learnt so much by observing the fusion of stellar black holes, how much can we deduce from the coalescence of two supermassive black holes, an event expected to occur when two galaxies are merging? But, in order to observe such a massive fusion, we must observe in a different frequency range, around a millihertz (mHz), and the detector must have the size of a few million kilometres. How can this be realized in space?

Early plans

Our story starts in the fall of 1974. Rainer Weiss chaired a committee where a NASA team presented a cross-shaped space interferometer, whose arms were 1 km long. Four masses of 1,000 kg were attached to the ends and the whole spacecraft weighed 16.4 tonnes (Figure 10.1)! The proposal was to build it in space thanks to the space station. This grandiose project, which was never realized, is a testimony to the pioneering spirits of the time: the Apollo programme was going full speed, 5 years after the first man on the Moon, 6 years before the inaugural flight of the space shuttle.

Peter Bender was a member of this committee at the time. Together with James Faller, his colleague from the Joint Institute for Laboratory Astrophysics (JILA) of the University of Colorado at Boulder, he had worked on tests of general relativity using the technique known as lunar laser ranging: they had convinced the Apollo 11 team to leave on the lunar surface a reflector, that made it possible to monitor (to the centimetre precision) the Earth–Moon distance, thanks to the reflection of a powerful laser beam emitted from Earth. This provided an excellent test of general relativity because it made it possible to compare very precisely the orbits of the Earth and of the Moon around the

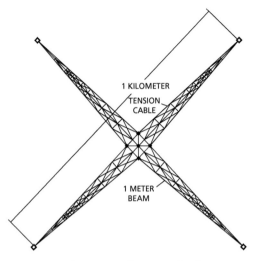

1 KILOMETER

TENSION
CABLE

1 METER
BEAM

Figure 10.1 First project of space interferometer for detecting gravitational waves (1974): we can recognize the structure of a Michelson interferometer.

Sun, and to check that they have identical accelerations, as predicted by general relativity.

I am stressing the Apollo connection because I do think that, more than 40 years later, the endeavour has retained some of the enthusiasm of this early period, where space was a symbol of Kennedy's New Frontier.

Coming back to Pete Bender, I met him some 30 years later, on the scientific team of the Laser Interferometer Space Antenna (LISA) project that gathered together both European and American scientists. Before realizing this was the famous 'Pete Bender', I had understood that he was someone special: very attentive to what was being said, informed about the smallest technical details as well as the broadest scientific aspects. Everyone turned to him when a complex question arose. Besides, he was a modest man, paying attention to newcomers such as I was. He is a great figure in our field.

Back in the late 1970s, Ray Weiss, Pete Bender, and James Faller, joined by Ron Drever, quickly realized that it was possible to locate the test masses in separate spacecraft, linked by lasers: space is a good enough vacuum to let the lasers propagate freely through it, outside any confined enclosure. Instead of the mirrors of a standard interferometer, it was possible to use transponders, which are relay devices that receive, amplify, and send back the laser beams: this technique makes it possible to increase significantly the distance between the satellites. In 1981, Bender and Faller proposed the Laser Antenna for Gravitational Radiation Observation in Space (LAGOS) mission. The basic concept was born (Figure 10.2): three satellites protecting freely floating test masses; laser beams link the satellites, making it possible to measure by interferometry the varying distances between two test masses; a distance between the satellites on the order of a million of kilometres; and an orbit around the Sun trailing the Earth. The project was well received by NASA but the hazards of funding prevented it from realization.

The baton was then passed to a group of European scientists, among which were Karsten Danzmann and Bernard Schutz in Germany, Stefano Vitale in Italy, and Alain Brillet and Pierre Touboul in France. The new project, named LISA for Laser Interferometer Space Antenna, was proposed to the European Space Agency (ESA) in 1993: it kept the three-satellite structure but the laser links now formed a triangle, making the spacecraft constellation fully symmetric (Figure 10.3).

The constellation, that is, the group of three satellites orbiting around the Sun, trails the Earth (in order to keep good communication contact with us) in such a way that the triangle slowly rotates without significant deformation (Figure 10.4).

If we were to start aligning numbers to define the performances required for the mission, it would make our heads spin. The three satellites must be a few million kilometres away (5 million in the original

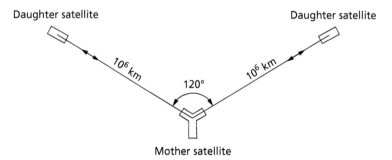

Figure 10.2 Original version of the LAGOS mission with a master spacecraft (with two test masses), and two daughter spacecraft (with a single test mass), and the laser links in between.

Figure 10.3 The LISA triangular constellation with its 3 satellites and the laser links, all orbiting around the Sun. © LISA Collaboration EAS/NASA.

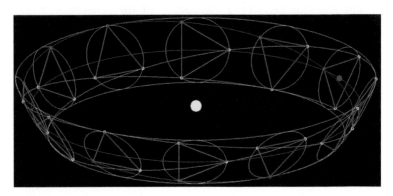

Figure 10.4 The LISA orbit around the Sun (not to scale). The orbit of each individual spacecraft is slightly destabilized with respect to Earth's orbit. © LISA Collaboration EAS/NASA.

proposal) and the precision that must be achieved in monitoring the varying distances between them is of order of the picometer, that is, 10^{-12} m. This sounds like science fiction.

Actually, physicists are not interested in measuring the *absolute* distance between two satellites, but only the *varying distances* between

A clever orbit

How do such clever orbits for the satellites of LISA get generated such that the triangle does not get distorted significantly as the satellites move around the Sun? Indeed, remember that there is no way to hold the three satellites besides gravity itself! To understand how one proceeds, just imagine putting a satellite along the orbit of the Earth around the Sun: it will trail the Earth on the same orbit around the Sun: both of them are freely falling in the gravitational field of the Sun and this is just another illustration of the universality of free fall that we discussed at the beginning of this book.

Now, slightly displace the satellite *away* from the orbit: you create an instability and the satellite will oscillate around the Earth orbit, sometimes above it, sometimes below it. Now, do this with three satellites cleverly disposed around the Earth's orbit, and you obtain the orbit of the LISA constellation (Figure 10.4).

them. In a way, it amounts to measuring how a mass moves a few picometers with respect to a reference, a task that is much easier than that faced by the physicists of LIGO and VIRGO who must reach precisions of 10^{-16} to 10^{-18} m!

Moreover, we are looking for waves, that is, periodic phenomena in the frequency range between 10^{-4} and 10^{-1} Hz. The period of such a wave is the inverse of its frequency: it is between 10,000 s and 10 s. In other words, we are only interested in phenomena that repeat themselves every 10 to 10,000 s. Any other periodic phenomenon does not spoil the measurement. For example, in the orbit around the Sun (see Figure 10.4), the triangle formed by the three satellites is slightly deformed: we say that the constellation is 'breathing', but this is not perturbing the search because this happens on time scales of a few months.

In the LISA concept, each satellite hosts two freely floating test masses that are associated respectively with the laser links to each of the other two satellites (Figure 10.5). Each pair of two arms forms an interferometer: thus, use can be made of three different interferometers. It takes light some 16 s to travel from one spacecraft to the next. Telescopes make it possible to focus the laser beam sent to the next satellite (as well as the beam received). Despite this, the light beam has a transverse size of some 20 km when it reaches the other spacecraft: most of the light is

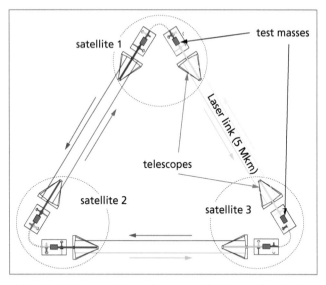

Figure 10.5 The LISA constellation (not to scale). We can identify the six test masses (two per satellite), the laser beams (one for each way) joining every couple of satellites, and the telescopes helping to focus them.

lost.[1] This is why light received is not reflected but re-emitted with identical properties.

With interferometry we are not actually measuring the precise distance between satellites, but the precise distance between the test masses that they host. Why this refinement? Well, you must remember that we want to measure a tiny gravitational effect. We thus want to make sure that we measure the distance between objects that are only subject to the gravitational force. If other stray forces acted on these objects, we might misinterpret their action for a passing gravitational wave! Despite the fact that space is a very quiet environment, the satellites are subject to many perturbations: solar wind, micrometeorites, cosmic rays, …

The real challenge of the LISA mission is to identify objects protected from these perturbations, and these are the test masses freely floating

---

[1] More precisely, if the power of the laser is 1 W at emission, only 10^{-12} W is received.

(a)

(b)

(c)

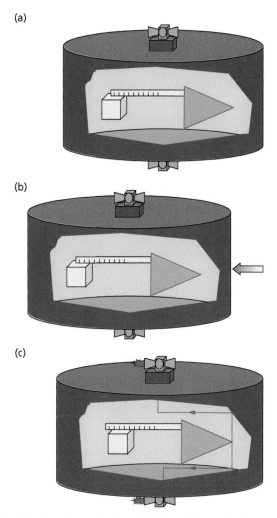

Figure 10.6 The drag-free technique. (a) The satellite includes an internal measuring device that makes it possible to identify its relative position with respect to the test mass. (b) The satellite is hit by external perturbations that modify this relative position. (c) When the effect becomes too important, and there is a danger that the mass will collide with its enclosure wall, microthrusters are briefly fired to reposition correctly the satellite around the test mass.

within their enclosures. And the satellites are used to protect them! The idea is that each satellite acts as a shield and absorbs the different perturbations to protect the test mass inside. The way this is achieved is by monitoring the relative displacement of the mass with respect to the satellite (or equivalently of the satellite with respect to the mass). Imagine that the satellite is hit by a micrometeorite (Figure 10.6): it moves slightly, whereas the freely floating test mass remains unperturbed. But the enclosure of the test mass moves with the satellite. Thus, seen from the satellite, it seems that the mass is moving in its enclosure. This relative motion is *detected* by the satellite. And the answer is to fire some very delicate microrockets, called microthrusters, outside the satellite, in order to reposition the satellite and the enclosure symmetrically around the test mass. This ingenuous scheme is called a drag-free technique.

These are very delicate manoeuvres and it was felt to be so central to the success of the mission that a technological mission was built to test this technique. This is the LISA Pathfinder mission of ESA, which was launched on 3 December 2015 (see Focus X). Six months later this mission splendidly confirmed the reliability of the technique, paving the way to the full LISA mission.

The science accessible to a gravitational wave space observatory

A space antenna like LISA operates in a frequency domain where extragalactic signals are accessible. Because the amount of matter involved is very large (several million solar masses), the signals have a large amplitude and most of them should clearly stand out from the background noise of the detector. The problem is thus not to extract the signal from the background, but rather to separate it from all the signals coming from the Universe. Indeed, millions of cosmic sources produce gravitational waves detectable in the LISA range. This is a real challenge for analysing data. The problem is comparable to that faced when one tries to reconstruct, from an audio recording in a stadium, individual conversations: one first identifies the loudest spectators (or those closest to the microphone), transcribes their conversations, and then suppresses them from the audio recording. And so on … Luckily, our own sources talk for months or even years, and have a rather limited and repetitive conversation!

In particular, one of the advantages of a space antenna is that its frequency domain corresponds to expected frequencies for sources already known: those are the millisecond binary pulsars that made it possible to confirm the existence of gravitational waves (Figure IX.4). The sources are called 'verification binaries' because the first task of LISA will be to identify them and to check that the gravitational waves they produce are in accordance with what has been observed until now. If this was not the case, this might cast some doubt on the performances of the instrument.

Among other astrophysical sources, it is clear that the most spectacular events are the fusions of supermassive black holes. We saw in Chapter 3 that the collision and subsequent merger of galaxies played an important role in the formation and structuration of the large regular galaxies that we know today: they were formed through a series of mergers of smaller galaxies (and accretion of mass in between). And this structuration continues: we expect that our Milky Way and Andromeda will merge into a single larger galaxy (Figure III.2). We have also seen that, at the centre of most galaxies, lies a supermassive black hole. A single one? In most cases, yes indeed. This implies that, during the merger of two galaxies, either one of them is kicked out or the two central black holes merge together.

Let us see in more detail how this merger of two black holes proceeds. The collision of the two galaxies leads to a spectacular shake-up of the stars that now feel the gravitational attraction of the other galaxy as well. The same occurs for the two central black holes. When the intergalactic storm calms down, the new galaxy starts shaping up and we meet again our two black holes rotating around each other.[2] The numerous gravitational interactions that they have had with the stars in the collision have slowed their rotational velocity, and now that the two black holes are close to one another, they can reproduce the elegant cosmic pas de deux, which we described earlier, in front of the dramatic backdrop of the two merged galaxies. They lose energy in the form of gravitational waves and get closer and closer until the climactic merge.

---

[2] Of course, gravitational interactions during the collision of galaxies could have acted as a sling and expelled one of the two black holes.

The difference with the same phenomenon with stellar mass black holes is that the phenomenon can be detected much earlier before the final plunge, because, being much more massive, the two black holes are emitting, even when they are inspiralling, a large energy in gravitational waves. An antenna like the LISA mission should be able to identify this energy. Figure 10.7 gives a more precise idea of this by comparing the amplitude of the signal with the instrumental sensitivity: the signal is detectable if it is above the sensitivity curve. We can note that in the case of two supermassive black holes (1 million solar masses) the signal could be identified approximately 1 year before the merger. This should leave ample time to predict the location in the sky and the precise time when the final event will occur. It is more than

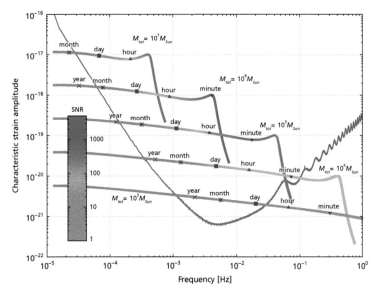

Figure 10.7 Amplitude of the signal during the final phases of the merger of two supermassive black holes (of identical mass, 10 million, 1 million, 100,000, 10,000, and 1,000 solar masses) one year, one month, one week, one day and one hour before the final plunge. The colour evolution from green (weak signal) to red (strong signal) shows the accumulated signal to noise ratio (SNR), i.e. the increase of signal power with observation time. The amplitude that the LISA mission can detect (called its sensitivity) is represented in red as a function of frequency: the amplitude of signals must be larger in order to be observable (courtesy: A. Petiteau).

probable that, for the first events announced (we can expect a few each year), the largest telescopes in the world will point towards this direction in the hope of detecting simultaneous electromagnetic signals!

Many other sources should be accessible to a gravitational space antenna. We expect, for example, a lot from the study of the fall of celestial objects into the gravitational field of a supermassive black hole: this would make it possible to map very precisely the geometry of space–time very close to a black hole's horizon.

These objects must be very compact; otherwise, they would probably be torn apart by the powerful tidal forces present near the central black hole's horizon. But a stellar black hole captured by the central black hole can orbit some hundred thousand times around the supermassive black hole's horizon while getting closer and closer until it vanishes into the horizon (Figures 10.8 and 10.9).

This provides a new way of mapping very precisely the geometry of space–time in the immediate vicinity of a galactic black hole's horizon. This should give us very precise information on the horizon of black holes and help us to answer a certain number of questions:

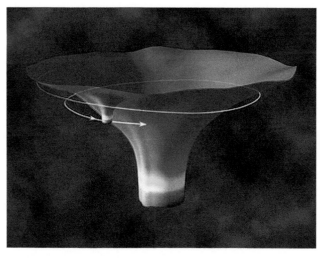

Figure 10.8 Stellar black hole orbiting around a supermassive black hole. The depth of the deformation represents the curvature of space–time at a given point. © NASA.

(a)

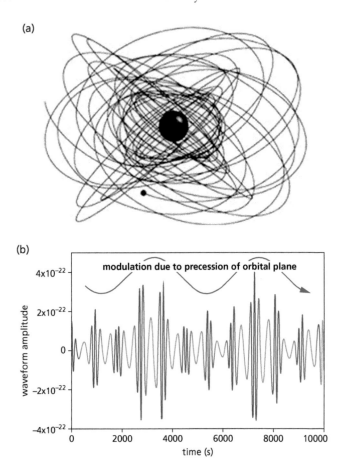

(b)

Figure 10.9 (a) Trajectory of a (not very massive) stellar-mass black hole around a very massive black hole before the final plunge (the total number of orbits is on the order of 100,000); (b) form of the gravitational wave emitted during this phase. © K. Danzmann et al, eLISA.

- What is the exact nature of a black hole's horizon? Is it totally opaque or does it leak information outside?
- What is the geometry of space–time very close to the horizon? Is it well described by classical general relativity? The question arises because the region close to the horizon is the location of Hawking radiation. Is it the only quantum effect or is quantum physics modifying in any significant way the geometry of the horizon?

- Is the black hole characterized by three hairs: mass, charge, and angular momentum? In other words, are astrophysical black holes satisfying the no-hair theorem?

It is fascinating to imagine that the space voyage to the horizon of a galactic black hole is no longer in the realm of science fiction but could be realized in little more than a decade thanks to a small black hole that would send us messages of what it observes, in the form of gravitational waves!

When the Universe was still opaque

Another advantage of gravitational waves is to give us direct information about the period before hydrogen recombination. The Universe was then filled with a plasma opaque to light ... but transparent to gravitational waves. Any violent event that happened during this primordial era generated gravitational waves that are still traveling through the Universe and can potentially be detected.

An example is provided by the phase transitions described in Chapter 4. When introducing the transition between the phase of quarks and gluons and the phase where matter is made of particles (protons, neutrons), I made an analogy with a well-known phase transition, the transition between liquid water and vapour. Let us observe a pot of boiling water to understand the dynamics of such a transition. At the beginning, only a few bubbles of gas (vapour) appear within the liquid water, then these bubbles multiply and collide with one another, forming pockets of gas that become increasingly large. This generates turbulence within the liquid. If we wait long enough, water disappears from the pot: it has been completely transformed into gas.

In the case of a phase transition in the primordial Universe, the nature of the vacuum changes as well as its energy. A new vacuum with a smaller energy, hence more stable, develops just like the gas within water. Bubbles of this new vacuum also develop within the old vacuum. They grow bigger and collide with one another, generating turbulence (Figure 10.10). Such phenomena (collision of bubbles, turbulence) generate gravitational waves. If they are sufficiently violent, the gravitational waves thus produced are in principle detectable by present or foreseen detectors.

Is the electric phase transition associated with the Higgs of this type, called first order? In other words, is it sufficiently violent to create detectable gravitational waves? If we follow the Standard Model strictly, the transition is smooth, but there are a certain number of arguments

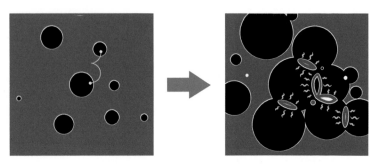

Figure 10.10 First-order phase transition. Bubbles of the new phase appear with the initial phase: they grow and collide with one another, thus generating turbulence.

that let us think that we should go beyond the Standard Model in the strict sense. In this case, it is absolutely possible that the electroweak phase transition is first order: a certain number of theories 'beyond the Standard Model' have this property. There would then be a remarkable convergence between experiments at very high-energy particle colliders and the search by LISA of gravitational waves coming from the most primordial times of our Universe.

LISA Pathfinder, green light for the full space mission

The LISA Pathfinder mission tests one key aspect of the future space gravitational wave observatory, known as LISA: it measures the variation, due to a passing gravitational wave, of distances between test masses that are in free fall, i.e. that follow purely gravitational trajectories. To protect the masses from any perturbation, there is a clever setup, known as 'drag free'. It consists in using the satellite for protecting the test mass placed in its centre from any perturbation.

To understand how it works, imagine that a micrometeorite hits the satellite: the satellite moves sideways, the test mass is thus no longer in its centre, the satellite detects this anomaly through sensors, it ignites some microthrusters to reposition itself around the test mass.

Of course, such a device is never perfect, and the test mass feels some tiny perturbations, but the goal of the LISA Pathfinder is to show that the perturbations are small enough that they still make it possible to have confidence in the detection of gravitational waves.

Quantitatively, the goal is to minimize the stray forces acting on the test masses. But a force induces acceleration and the goal of LISA Pathfinder is to minimize the stray acceleration (we talk of an acceleration noise). The initial goal of the mission was to reach over periods of 1,000 s

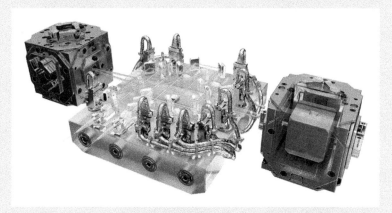

Figure X.1 LISA Technology Package core assembly without the vacuum enclosures. This is an artist's impression. © ESA/ATG medialab.

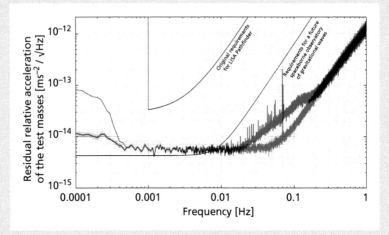

Figure X.2 LISA Pathfinder results. The graph shows the residual relative acceleration of the two test masses as a function of frequency. © ESA/LISA Pathfinder Collaboration.

Figure X.3 Lisa Pathfinder satellite. In the assemblage room when it was placed in the Vega launcher. © S. Martin, ESA-CNES-Arianespace.

a stray acceleration that is smaller than 10^{-13} times g, the local acceleration due to Earth's gravity.

How to realize this? In the future LISA mission, the test masses are placed at the centre of each satellite, 5 million km apart. Laser beams connect the three satellites, thus forming a huge triangle. Relative varying distances are measured by interferometry, just as for ground detectors, such as LIGO.

In LISA Pathfinder, one arm of the future LISA mission is reduced to 38 cm in order to locate two test masses in a single spacecraft. The distance between these two masses is monitored by laser beams that form an interferometer very similar to that on board LISA (apart from the distance covered by the beams) (Figure X.1).

It is not possible to have simultaneously the two masses in free fall, because their orbits are very similar but not exactly identical. This is why one of the two masses is used as a reference, whereas the other one is left free. It is with this second mass that we check whether it is in free fall, at least in the limits required for the acceleration noise.

This technological mission was launched on 3 December 2015 from Kourou in French Guyana. The first results of the LISA Pathfinder mission were presented in June 2016, six months later. And they were even better than was anticipated.

'The measurements have exceeded our most optimistic expectations,' says Paul McNamara, LISA Pathfinder Project Scientist. 'We reached the level of precision originally required for LISA Pathfinder within the first day, and so we spent the following weeks improving the results a factor of five better.'

Indeed, as shown in Figure X.2, which appears in the published paper of the ESA/LISA Pathfinder collaboration, the acceleration noise reached is five times smaller than what was required; basically, it is already what is needed for the LISA mission, and even better for high frequencies.

'Not only do we see the test masses as almost motionless, but we have identified, with unprecedented precision, most of the remaining tiny forces disturbing them,' explains Stefano Vitale, the scientist in charge of the mission (Principal Investigator).

This success is obviously a green light for the gravitational wave observatory, the third large mission (L3) of the European Space Agency, known as LISA. This mission was originally identified for a launch in 2034 but this success, and the historic discovery of gravitational waves by the LIGO detector, offer strong arguments to advance significantly the schedule (Figure X.3).

11

Epilogue:
Towards What Future?

So far we have tried to *predict* the past evolution of our Universe. This obvious oxymoron, in fact, summarizes our scientific approach: we use today's observations of our Universe to make predictions about some of its currently unobserved properties, and then we verify that they are indeed visible in the sky, which after all contains information about our past. This begs the following question: does it make sense scientifically to make predictions about the future of our Universe, while we will not have the possibility of verifying those predictions?

The debate is open, and it can be addressed at different scales.

In our immediate neighborhood—that is in the Solar System—the trajectories of planets and other bodies, which surround us, are sufficiently well understood that we can predict their future positions with high precision.

On larger scales, of the size of our own Galaxy or even further out to extragalactic scales, we can start from the postulate that we do not occupy a privileged position in space–time. As a result, all the events that we observe in the Universe could already have happened to us— or could happen in the future—under the right conditions. For instance, in the sky we observe collisions of galaxies, at different stages of evolution. Astrophysicists therefore conclude that our own Galaxy could have gone through a similar process and can continue to do so in the future.

From observations of the galaxy the closest to us—Andromeda— such as its velocity relative to us, we can deduce that it is probable that in the future our Galaxy will collide with Andromeda.

On the largest of scales, that is, of the size of our own Universe, we do not have enough observational data. One way around that is to imagine subdividing the Universe into smaller parts. Say, for the sake of argument, that we divide the observable Universe into 100 samples, and using them we carry out a statistical analysis to determine the averaged properties of each sample. Indeed, this is precisely the way we estimate that fluctuations in cosmic microwave background (CMB) temperature are of order 1 in 100,000. However, had we divided the observable Universe into only 4 regions, then our statistical analysis would have been based on only those 4 samples and it's extremely likely that the resulting observed CMB temperature fluctuations would be large. This effect is known as *cosmic variance*. An intuitive understanding of cosmic variance can be obtained by looking at the CMB map (Figure 5.9): divide it into 4 equal parts and compare the maps in each dial. You will observe that these maps are not similar! However, don't be tempted to reach a hasty conclusion from this simple analysis—all experimentalists who have measured the same quantity only 4 times have observed big statistical fluctuations between the measurements!

So, can we make predictions about the future of our Universe? The press has come to its own conclusion: the easiest way for a scientist to get press coverage is to predict some catastrophe or other! And if it is on the scale of the Universe, so much the better!

Though it's risky to make predictions, nonetheless it's a tempting game to play intellectually. Therefore, please allow me to do it in this last chapter. In an attempt for your forgiveness, towards the end of this chapter I will come back down to earth and carry out a simpler exercise: we will outline the different observational techniques and experiments being developed today, which should enable us to finally determine directly the gravitational aspects of our Universe in the next 20 years.

Predicting the future of the Universe?

In the relatively recent history of the Universe, a new form of energy—known as dark energy—has started to dominate. This dark energy seems to have very little dynamics. Indeed, if it is vacuum energy (as seems likely), its value will be constant in time (unless there is a transition phase). This should be contrasted with other forms of energy (such as mass and radiation) whose values diminish with time.

As a result, dark energy should become more and more dominant, and thus lead to continuous accelerated expansion of our Universe. This is similar to the behaviour of the Universe during the inflationary phase.

This new phase of inflation will, in the long term, accelerate the expansion of the Universe to such a degree that large-scale structures, such as clusters of galaxies and even galaxies themselves, will start to break apart. (This, despite the fact that, as we have seen, these structures are, to a good part, protected from the expansion of the Universe since they are gravitationally bound systems.) Then it will be the turn of smaller structures such as stars and their planets. Finally, all material structures will split into particles which themselves will move apart from each other.

We can only hope that, as in the inflationary phase, the Universe will fall into another quantum vacuum state of lower energy. If so, the energy freed in this process will allow the creation of particles, followed by atoms, molecules, and then larger and larger structures.

At this stage, we can wonder if our observable Universe is perhaps just one of many other Universes? This 'multiverse' hypothesis was first put forward by Andrei Linde, and it was, in fact, partly motivated by the inflationary scenario itself.

The idea is one in which inflation occurs eternally. This works as follows: in some region of a Universe, which itself was formed in an inflationary phase, a quantum fluctuation of vacuum energy can trigger inflationary expansion; when that happens, the local region starts expanding exponentially and quickly forms a new Universe, causally disconnected from the previous one (apart from the region from which it was born). And the process carries on. We therefore end up with a distribution of Universes, called multiverses, represented by Linde by the diagram shown in Figure 11.1. Each Universe can have very different properties, for instance, different numbers of spatial dimensions, different types of particles, different masses for these particles, etc. Theoretical physicists then try to calculate the probability that we could live in a Universe with the actual properties which we observe in our own Universe. In particular, this type of calculation—similar to the anthropic argument mentioned at the end of Chapter 6—can be applied to try to explain the value of the observed vacuum energy density. Furthermore, we can try to explain other parameters of our Universe, such as the mass of the Higgs or the ratio of masses of the electron and proton.

This type of approach, however, must overcome an important difficulty, namely that it is difficult to define a probability distribution on the ensemble of all the Universes in a rigorous mathematical way. Is this a temporary problem, which will be overcome in due course, or one which is intrinsic to this type of approach? Only the future will tell.

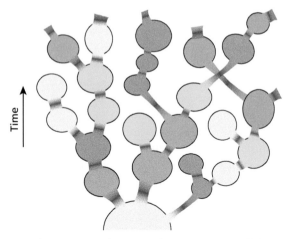

Time

Figure 11.1 Multiverses according to Andrei Linde. Each bubble represents a Universe.

That said, for many theoretical physicists, eternal inflation is an absurdity of the inflationary scenario—something which simply suggests that inflation is an incomplete description of the physical world. For others (I do not belong to this group!), the 'probabilistic' calculation of a certain number of parameters, such as vacuum energy, presents a new paradigm, which, furthermore, may be supported by string theory. Whatever the conclusion, in cosmology conferences, this type of discussion leads to interesting verbal sparring matches … often featuring personal attacks. And often it seems that beyond the scientific arguments, there is a clash of different personal philosophical opinions.

The end of time?

Some researchers suggest that one way to resolve the problems of eternal inflation is just to keep a fraction of the multiverse. But in that case, certain observers (us?) would appear to reach the boundary of this portion of multiverse in a finite time. In other words, in such a scenario (which is of course highly speculative) time would only have a finite duration: it would come to an end. Or rather, it should be replaced by another concept. Should that come as a surprise? Take another look at the evolution of our Universe summarized in Figure 4.2, and note that there is a striking symmetry between the early inflationary phase and the late time accelerated phase dominated by dark energy. If this symmetry is not pure coincidence, we can envisage the idea that our Universe is evolving towards a phase very similar to the one that preceded inflation: that is, one in which the concept of continuous space and time disappear.

The observer

I have mentioned many times that the observer seems to play a central role. That might appear to be incompatible with the idea that there is no privileged position in the Universe. However, the very fact of recognizing the privileged role of the observer means that we also recognize the essential and thus privileged role of observational data—or, to be more precise, of physical measurements.

Indeed, quantum mechanics (both relativistic and nonrelativistic) has taught us that the very act of making a measurement is essential in order to obtain accurate predictions and coherent results. Regarding this point, I will never forget the following question, repeated so often by Raymond Stora, a great specialist of quantum field theory: 'What are the observables?' If, in quantum field theory, you try to calculate the value of a quantity that is not observable, there is a high chance you'll obtain an absurd answer, such as infinity. It is precisely in order to tackle this type of problem that the theory of renormalization was developed by Feynman, Schwinger, and their colleagues in the 1950s.

It therefore appears that to tackle head on the question of how to reconcile quantum theory and general relativity, it is necessary to return the observer to her privileged status, and recognize her central role in the process of measurement. Of course, any other observer is equally privileged, and any other measurement just as important!

Speculating even further, it seems that even the notion of the Universe's evolution, and therefore of time itself, is directly related to the act of measurement. In other words, without an observer, the Universe would be static: it is the act of measurement which gives rise to the evolution. Of course, this is not true for the individual measurement that I am making myself, at this very minute, with my telescope! But it would be true of a more global measurement shared by all observers: it is this one which would define the evolution of our Universe. Finally, let me remind you that we can define the content of the Universe in terms of information. The question of the Universe's evolution can therefore be rephrased in terms of an information flux.

Let us finish with speculations though! One of my aims in outlining some of them was to show you that physicists also know how to put their equations to one side and search in the dark for a path towards a solution. Once that path is found, it's necessary to construct a mathematical formalism, a theory, in order to back up the intuitions. And then, finally, we must test the theory with observations. Let us therefore end this last chapter by presenting some of the most important experimental developments of the most recent years and the next decade or two: these should, in due course, bin all the crazy theories whilst putting the final flourishes on the new theories of tomorrow.

Will the next two decades be gravitational?

It goes without saying that the experiments being developed at the moment to test the theory of gravitation more than do justice to the anniversary of 100 years of general relativity. The gravitational force, weakest of all the fundamental forces, should (like Cindarella!) finally have its time in the spotlight! Indeed, very large and costly experiments, both on Earth and in space, are being developed. Of course, we have already discussed gravitational wave detectors, which will open up a window onto the gravitational Universe.

But other experiments to test general relativity itself, in the finest of details, are being developed. Let me mention, for instance, the so-called 'Microscope' mission (name derived from the French 'Micro-Satellite à traînée Compensée pour l'Observation du Principe d'Equivalence'), which started running in 2016. Microscope will test the equivalence principle, namely the equivalence between inertial and gravitational masses, to a precision of 10^{-15} (Figure 11.2).

More precisely, the aim is to test one of the consequences of the equivalence principle (see Chapter 1): the fact that the motion of a body in a gravitational field depends solely on its mass, and not on its internal composition. Now, it so happens that in a certain number of theories that try to unify gravity with the other fundamental forces, one part of the rest-mass energy ($E = mc^2$) is sourced by new interactions that depend on the composition of the body. In that case two bodies of the same mass but of different composition would fall differently: the

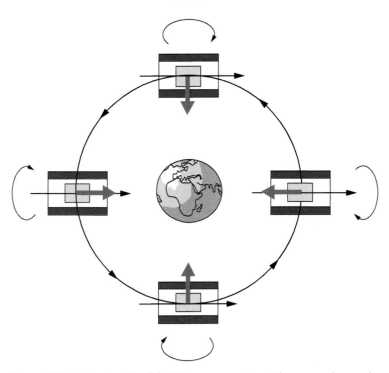

Figure 11.2 Basic principle of the Microscope mission. The arrows denote the electrostatic force required to keep the internal mass stationary relative to the external mass (two concentric cylinders). The whole system is in orbit, and also rotating, the idea being that in that way it is easier to identify possible violations of the equivalence principle as opposed to random perturbations.

universality of free fall, which for the moment is verified to an accuracy of 10^{-13}, would be violated.

In the way it is set up, Microscope has some similarities with LISA Pathfinder; namely, microthrusters are used to compensate for disturbing forces. However, this is done in a slightly different way to LISA. The test masses are two concentric cylinders of the same mass but of different composition (platinum and titanium). Using electrostatic forces, these masses are kept immobile inside a cage made of silicia.

If it were to find any evidence of violation of the equivalence principle, then it would be a clear sign either that general relativity must be modified or that it is necessary to introduce a new long-range force which, contrary to general relativity, depends specifically on the composition of a body. Thus, at the degree of precision of Microscope, we could learn either that the gravitational force behaves differently or that there exists another fundamental force. Such a result would give hints on how gravity can be unified with the other fundamental forces.

Confirmation of Einstein's theory of relativity by Microscope

On December 4, 2017 the Microscope mission unveiled its first results, which confirm with the unprecedented precision of 2.10-14 the universality of free fall. The equivalence principle remains unviolated for now. The Microscope satellite has gathered data from 1,900 orbits, the equivalent of 85 million kilometres in free fall (half the Earth–Sun distance). We expect these first results will soon reach the precision of 2.10-15.

In this book, I have only rarely broached the subject of whether these big fundamental questions will change our everyday life in the immediate future. However, it's an often-posed question. One illustration of the utility of general relativity is in GPS systems: to localize one's position on Earth to within a few meters, a message (electromagnetic wave) must be sent to the GPS (or Galileo) constellation at 20,000 km altitude. Using a very precise measurement of the arrival time of the message, it is possible to locate the position of the person using the triangulation method we have already discussed. But we know that measurements of time and spatial distances are subject to distortions due to relativistic effects. These distortions are tiny, but not negligible if we want to have an accuracy of 1 m in our measured position on Earth. (For a satellite at 20,000 km altitude—or even more if the satellite is not exactly overhead—that requires a precision of 10^{-7}). Indeed, for such an accuracy, corrections due to general relativity are not negligible and they must be taken into account. So, every time we use a GPS to localize our position, we are using general relativity!

In the future, our everyday life is likely to become more and more gravitational. Already now, the satellites GRACE in the USA and GOCE in Europe are measuring Earth's gravitational field with higher and higher precision, so as to explore not only the details of terrain (both continental and marine), but also Earth's crust. Usually the assumption is that Earth's gravitational field is the same at all points on its surface. However, this surface has relief, and furthermore the mass of the crust depends on its composition. And finally, the thickness of this crust varies over the Earth. For all of these reasons, the gravitational field at Earth's surface is not constant, and a precise measurement of its variations gives crucial information about the structure of Earth's external layers.

The GOCE experiment of the ESA, launched in 2009 and terminated in 2013, measured with exquisite precision the motion of an ensemble of test masses due to Earth's gravitational field. This was done using the mechanism of compensation already discussed in the concept of LISA Pathfinder and Microscope. By measuring tiny variations between their respective motions, it was possible to map the gravitational field of the region just below the passing satellite. In this way, we now have very useful maps of Earth's geoid—namely the surface of constant gravitational field (Figure 11.3).

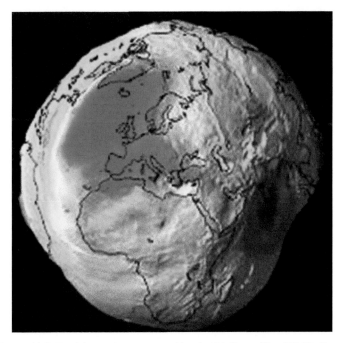

Figure 11.3 Earth's geoid as measured by the EAS's satellite GOCE. © ESA.

To conclude, an impressive experimental and observational programme has been developed to explore the gravitational Universe.
Clearly, gravitational wave detectors play a central role. How many
sources will be observed per year? Of which type? Will there be unexpected sources? Or will the study of expected sources give new sur-
prises? Only the future will tell. This exploration of the gravitational
Universe is particularly important as it is linked to some of the most
fundamental questions we can ask about our place in the Universe, the
nature of space and time, and even of the origin of the Universe and its
future. What an exciting outlook for a human being who passes the
first year of life trying to understand how to deal with gravity!

A question of time

Time is invention or it is nothing at all.

HENRI BERGSON,
Duration and Simultaneity (1922)

Since I've mentioned the future, now seems like an opportune moment to discuss time itself. Time has imposed itself as a central concept in contemporary physics, but it comes in different guises, which, at this point, are important to distinguish. Let's straightaway put to one side psychological time. Even if time appears to go at different rates for different people, and on different occasions, this effect has nothing to do with physical time. The well-known effect of time dilation, which has to do with moving frames (see Focus I), is a physical effect: take two clocks, which are initially synchronized. The one that travelled by plane will be slowed down as compared to the other clock, which stayed on Earth. This is not because the clock got bored, but because physical time was longer!

Physical time is directly linked to causality—namely the past, the present, and the future—and only events which occurred in the past can influence present or future events. It's actually for that reason that it's difficult to conceive the idea of more than one time dimension. Indeed, if time was a 2-dimensional surface, we could draw a circle on that surface: on going around the circle we would find ourselves in the past of our future! For instance, going back to the example already mentioned in Chapter 5, the fact that my milk boiled over this morning could influence, and even be the cause of, the death of Molière! That would be a good subject for a science-fiction detective thriller, but it's hardly compatible with physical reality.

In certain extreme cases, it's space, which can have a direction. For instance, a particle which goes through the black hole horizon will necessarily always fall into the singularity at the center of the black hole: an event situated at two-thirds of the Schwarzschild singularity will always occur before one at half the radius. Space drapes itself with the attires of time.

Often much emphasis is placed on the fact that the laws of fundamental physics are reversible: that means that the same laws would apply if the direction of time were reversed. For example, suppose I drop a ball. If I asked my favorite wizard to press the 'time-reverse' button, then the ball would travel back up into my hand according to Newton's laws. This may seem obvious, but it no longer holds when many particles are present: if I open a capsule containing gas in a room, the gas will spread into all the room. Reversing time will not make all the gas molecules enter the capsule again. It's only in fairytales that good and bad spirits go back into the bottles from which they were released! This is the essence of what is known as the second law of thermodynamics: many-body systems evolve from order to disorder. Entropy is precisely the quantity that measures disorder: a system always evolves in time such that entropy increases. This is irreversible.

What about the gravitational collapse of a star into a black hole? Is that process really reversible? Is it fundamental? If we assume that the final object, the black hole, is relatively simple, then the answer is in principle yes. On the other hand, if the collapse involves many particles, then these will lose their individual nature.

From this point of view, it shouldn't come as a big surprise that Stephen Hawking was able to define an entropy for black holes, and that he managed to show that this entropy obeys the same laws as standard entropy: it increases with time. Black hole entropy is measured by the area of the horizon, which is yet another sign that the horizon plays a central role in these different questions. When a physical object falls into a black hole, the mass of the black hole increases, hence the area of horizon increases, and as a result so does the entropy.

During this process, information (on the nature of the object, its properties, its mass, etc.) has been absorbed by the black hole. This information is directly related to the notion of entropy, and hence it is important to follow the information flow in order to understand what is going on. Indeed, the black hole will end up evaporating by emitting Hawking radiation. Will that release all the information which it absorbed? We can ask the question a different way: will black hole entropy be returned in the form of usual entropy? These are some of the types of questions that the scientific community is addressing.

So we see that the process of gravitational collapse and the formation of a black hole, so specific to gravitation, is probably what makes gravity stand out from the other fundamental forces. It's for this reason that hints regarding answers to fundamental questions related to time are interlinked with horizons of black holes. And these hints are now per-haps within the reach of our experiments!

Finally, if there's a link between the black hole horizon and the cosmological horizon, as certain approaches based on the holographic principle suggest, then the nature of cosmological time itself could come to light thanks to those experiments. And then we may learn whether the continuous time we know has a beginning and an end, and if so what quantity would replace it.

To Go Further

Some references from initiation books to comprehensive treatises

Galileo Galilei, *Dialogues Concerning Two New Sciences* (1638).

George Gamow (and Russell Stannard), *The New World of Mr Tompkins*. Cambridge University Press, Cambridge (2001).

Anthony Levi, *Blaise Pascal, Pensees and Other Writings*. Oxford World's Classics, Oxford (2008).

François Cheng, *Empty and Full: The Language of Chinese Painting*. Shambhala, Boston & London (1991).

Edwin A. Abbott, *Flatland: A Romance of Many Dimensions*. Dover Thrift Editions, New York (1992).

Stephen Hawking, *A Brief History of Time*. Bantam Book, New-York (1988).

Additions on relativity

Marc-Antoine Mathieu, *3"*. Jonathan Cape, London (2013).

Two classical on black holes

Kip Thorne, *Black Holes and Time Warps: Einstein's Outrageous Legacy*. W.W. Norton, New York (1994).

Jean-Pierre Luminet, *Black Holes*. Cambridge University Press, Cambridge (1992).

A vivid and well-informed account of the quest for gravitational waves

Marcia Bartusiak, *Einstein's Unfinished Symphony*. Berkley Books, New York (2000).

For those of you who had some problems understanding the aspects of quantum physics approached in this book

Thibault Damour and Mathieu Burniat, *Mysteries of the Quantum Universe*. Particular Books, New York (2017).

On cult movies

The Incredible Shrinking Man, Director: Jack Arnold, Scriptwriter: Richard Matheson.

Kip Thorne and Christopher Nolan, *The Science of Interstellar*. W. W. Norton, New York (2014). This book describes the science behind the movie; Kip Throne is an eminent scientist who specializes in black holes.

Glossary

ACCELERATION The rate of increase of the velocity of an object. If the acceleration is zero then the velocity of the object is constant this is called: uniform motion. If the acceleration is constant, the object is said to be uniformly accelerating.

ACCELEROMETER A sensor attached to an object which measures the acceleration of that object.

ANGULAR MOMENTUM Quantifies the rotational state of a physical system.

ANTIMATTER Matter consisting of antiparticles. Most particles have associated antiparticles, of the same mass (m) but of opposite charge. When particles and antiparticles annihilate each other, an energy equal to $2mc^2$ is released in the form of photons or other particles. Some particles, such as photons, are their own antiparticles.

ATOMIC TRANSITION Motion of an atomic electron from one energy level to another.

BARYON Particles formed by three quarks: the most common examples are the proton and neutron. Baryonic matter, i.e. matter formed by baryons, is typically ordinary matter formed by protons and neutrons.

BARYONIC MATTER Matter formed by baryons (typically protons and neutrons). This expression is used in a broad sense to refer to all potentially luminous matter, as opposed to dark matter.

BLACK BODY An object that absorbs all electromagnetic radiation. It does not reflect any light and in principle appears black. At nonzero temperature, however, it emits radiation which has properties characteristic of that temperature. This is called black body radiation, the origin of which was explained by the physicist Max Planck in 1901.

BLACK HOLE An astronomical object that is sufficiently compact (namely, sufficiently massive and dense) it can trap both matter and light in its vicinity.

BOLOMETER Detector that measures the energy stored in an electromagnetic field by converting it to heat.

BOSON Type of particle characterized by the following property (as opposed to fermions): an arbitrary number of bosons may be in the same microscopic state and form a coherent macroscopic state. Since the photon is a boson, laser light is a coherent superposition of these photons. In general, the bosons are the mediators of forces.

CASIMIR EFFECT Attractive force exerted between two uncharged conductive plates in vacuum. It is due to the fluctuations of the electromagnetic field in vacuum.

CLOSED (SPACE) Finite space with or without a boundary for example, the surface of a sphere does not have an edge (unlike a flat or open space).

CLUSTER OF GALAXIES Set of more than hundreds of galaxies bound by the force of gravity.

COSMIC MICROWAVE BACKGROUND Electromagnetic radiation produced by the hot and opaque Universe (comparable to a 'black body') 380,000 years after the Big Bang, at the precise moment when the Universe became transparent. This primordial radiation has undergone a spectral shift and is now observed in the microwave domain. It is sometimes referred to as the 'first light', which is not quite exact since light (i.e. electromagnetic radiation) was produced in earlier periods; however, it was immediately captured by matter and could not propagate.

CRITICAL ENERGY DENSITY Average energy density in the Universe corresponding to a flat space: the precise value associated with the current observations is of 10^{-26} kg/m^3. If the average density in the Universe is above this value, the space is closed; if it is below, the space is open.

CURVATURE (OF SPACE–TIME) Quantity measuring how far a geometrical object is from being flat. For example, the radius of a sphere measures its nonflat character the smaller the radius, the more curved the sphere.

DARK ENERGY Form of energy responsible for the accelerated expansion of the Universe observed in its recent history. Its nature is yet to be determined.

DARK MATTER Non-baryonic form of matter which constitutes most of the mass of galaxies and galaxy clusters, as well as most of the matter in the Universe. It is dubbed 'dark' because, as opposed to baryonic matter, it does not emit light.

DIMENSION (i) Quantity measuring the size of an object in a given direction (length, width, height). It is in this respect that we say that time is the fourth dimension, or that we talk about higher dimensions which correspond to dimensions not directly noticeable in our perceptible world. (ii) Product or ratio of fundamental physical quantities (mass, time, length) from which other physical quantities can then be derived hence, force is the product of mass by length divided by squared time. The corresponding analysis of the derived quantities is called dimensional analysis. In the International System of Units (sometimes called metre–kilogram–second–ampere), this makes it possible to obtain a derived unit from the fundamental units: hence the Newton, a unit of force, is given as kg.m/s^2.

DOPPLER–FIZEAU EFFECT Effect on the frequency (or the wavelength) of a wave emitted by a moving body and detected by a motionless observer the detected frequency is greater (the sound is more high-pitched in the case of sound waves) if the body is approaching, and smaller if it is receding (the sound is deeper).

DRAG A force that acts in the opposite direction to an object's motion in a liquid or gas. The mechanism of drag compensation applied to a satellite corresponds to performing an action on the satellite (using some microthrusters) in order to correct the effect of drag on the trajectory (allowing the mass located in the satellite to move under the sole effect of gravitational force).

ELECTROMAGNETIC SPECTRUM Distribution of electromagnetic waves as a function of their wavelength, or equivalently of their frequency (which is directly given by the speed of light divided by the wavelength). For increasing wavelength (decreasing frequency) we respectively have gamma rays (from 10^{-14} to 10^{-12} m), X rays (from 10^{-12} to 10^{-8} m), ultraviolet radiation (from 10^{-8} to 4×10^{-7} m), visible radiation (from 4×10^{-7} to 8×10^{-7} m, namely from blue to red, see the figure in the Chapter 3 box 'Spectral Redshift'), infrared radiation (from 8×10^{-7} m to 1 mm), microwaves (from 1 mm to 300 mm) and radio waves (from 300 mm to thousands of km).

ELECTROMAGNETIC WAVE (see also LIGHT WAVE) A wave which corresponds to the motion of periodic electric and magnetic fields (the motion of an electric field creates a magnetic field and reciprocally). This wave has an equivalent description in terms of photons, the quanta of light. An electromagnetic wave is generally referred to as 'light', whereas the term 'light wave' is more specifically reserved to visible light.

ELECTRONVOLT Unit of energy: one electronvolt (denoted by eV) is the kinetic energy acquired by an electron in an electric circuit powered by a 1-volt battery. It is a microscopic unit; hence, gigaelectronvolt (denoted by GeV) is often used to denote a billion electronvolts.

ENERGY DENSITY Ratio between the total energy contained in a given region of the Universe and the volume of this region: the more concentrated the energy, the greater the energy density.

EQUATION OF STATE Relation between physical parameters (pressure, temperature, energy density, etc.) that determines the state of a system. In cosmology, this system is the entire Universe at a given time (today, during hydrogen recombination, during inflation, etc.).

EQUIVALENCE PRINCIPLE Principle establishing an equivalence between acceleration and a gravitational field: the two phenomena *locally* have the same physical consequences. One of the consequences of this principle is the equality between mass as a measure of inertia (hence, associated with acceleration) and the mass appearing in Newton's law of universal gravitation (associated with gravitation).

EXPANSION OF THE UNIVERSE Expansion of the very fabric of the Universe that causes the galaxies to move away from each other over time (the opposite movement corresponds to the contraction of the Universe).

EXTRAGALACTIC All the objects that are not within our Galaxy (Milky Way).

FERMION Type of particles that obey Pauli principles (unlike bosons): two particles cannot stay in the same microscopic state. Typically matter particles are fermions.

FIELD In physics, a field is a quantity that has a value for each point in space. Examples are a pressure or velocity field in a fluid, or an electromagnetic field. If the quantity is a number (e.g. the pressure), it is called a scalar field. If the field is a vector (e.g. the velocity field or the electric field), it is called a vector field.

FLAT (SPACE) A flat surface is a surface with no relief. It becomes more difficult to define without any mathematical formalism what flat space is in geometry: let us just indicate that it is a space where the sum of the angles of a triangle equals 180°, and where parallel lines never cross (according to the rules established by Euclid).

FORCE An action capable of setting an object in motion or of changing the movement of an object or of any of its parts (and thus eventually deforming it).

FREQUENCY Number of times per unit of time at which a periodic phenomenon is identically reproduced. In particular, we consider the frequency of a wave. The unit of frequency is hertz (short: Hz): A frequency of 1 Hz corresponds to a phenomenon which occurs once every second. The inverse of the frequency of a wave is called its period (measured in seconds).

FUNDAMENTAL FORCE A force described by a law which takes the same form at any point in time or space. We only know four fundamental forces or interactions electromagnetic, weak, strong, and … gravitational.

FUNDAMENTAL INTERACTION Synonymous to fundamental force but indicating that it is an interaction between two bodies (for example, two masses for the gravitational interaction or two charges for the electromagnetic interaction). At a quantum level, this interaction is due to the exchange of fundamental mediator particles of the interaction between the two bodies (the photon for the electromagnetic interaction, the gluon for the strong interaction, the W, Z, and Higgs bosons for the weak interaction, and … the graviton for the gravitational interaction).

GALAXY The sum of dark matter, stars, gas, and dust, which is held together by its gravitational interaction. A galaxy like ours (the Milky Way) has a size of about 100,000 light years and contains a few 100 billion stars. There are three main types of galaxies: elliptical ones of spheroidal form (old stars, only little gas and dust), spiral ones like the Milky Way (relevant quantities of gas and dust, young stars in the disk, old ones in the bulk), and irregular ones (of smaller size and rich in young stars).

GALILEAN (OR INERTIAL) REFERENCE FRAME Reference frame for which the law of inertia is applicable an object preserves its status of uniform motion unless it is acted upon by a force. These reference frames are *uniformly* moving with respect to one another. (See also REFERENCE FRAME)

GEV (see ELECTRONVOLT)

GRAVITATIONAL FIELD (OR FIELD OF GRAVITATIONAL ATTRACTION) Region of space where a gravitational force works or *alternatively* the value of the gravitational force acting on a unit mass placed at a given point.

GRAVITATIONAL WAVE A wave of space–time curvature creating a perturbation in the measure of distances. This perturbation travels at the speed of light.

GRAVITON Elementary particle whose exchange between particles is responsible for their gravitational interaction. The role of the graviton in the gravitational force is analogous to the role of the photon in the electromagnetic force.

HAWKING RADIATION Stream of particles produced near the horizon of a black hole, due to quantum phenomena (formation of pairs of virtual particles, one member of the pair then disappearing into the horizon). This production leads to an energy loss of the black hole, which is known as black hole evaporation. The progressive evaporation of the black hole eventually leads to its disappearance.

HOMOGENEOUS Composed of elements or parts with similar properties.

HORIZON Border of the region of space–time that an observer can see, or more generally that he can probe with any type of detector. The horizon of an event is defined as the border of the region that an observer could explore if he waited an infinite amount of time (or, equivalently, of the region that he can never observe). In contrast, the past horizon is the border of the region from which he could have, in the past (since the Big Bang), sent a signal.

HUBBLE (LAW, CONSTANT, PARAMETER) According to Hubble's law galaxies move away from us with a speed proportional to their distance to us. The constant of proportionality is the Hubble constant. It also corresponds to the expansion rate measured today. Since expansion has changed over the history of the Universe, the cosmological expansion rate is a function of time, called the Hubble parameter, whose present value coincides with the Hubble constant.

INERTIA Quantity characterizing the resistance of a body against changes in its motion (for example, change of its speed). The inertia of a body is proportional to the amount of material it is made of. The mass of a body is a measure of its inertia.

INFLATION Period in the history of the Universe, probably just after the Big Bang, where expansion was extremely (exponentially) fast.

INTERFERENCE Phenomenon resulting from the superposition of two waves of the same nature with identical or similar frequencies, which manifests itself by a variation in space (and time) of the amplitude of the resulting wave. At certain points, the interference is constructive and the resulting wave has enhanced amplitude, while at other points the interference is destructive and the resulting amplitude is weaker or even vanishing. In the case of light waves, this leads to a system of alternating dark and light patterns.

INTERFEROMETER (see MICHELSON INTERFEROMETER)

ISOTROPIC Showing the same properties in all directions.

KELVIN Unit of temperature (symbol: K) used in the International System of Units. The Kelvin represents an absolute measure of temperature: its zero corresponds to absolute zero, which is the configuration where thermal motion of molecules stops. The Kelvin was chosen such that a variation of 1 K corresponds to a variation of 1°C. Hence we have the following: 0 K= −273.15°C, 273.15 K = 0°C, 373.15 K = 100°C, etc.

LIGHT WAVE (SEE ELECTROMAGNETIC WAVE) Electromagnetic wave whose wavelength is in the visible spectrum, that is, between 0.38 and 0.78 microns (corresponding to photons of energy between 1.5 and 3 eV).

LIGHT YEAR Distance travelled by light in one year, which corresponds to 9460 billion km.

LUMINOSITY Total amount of energy radiated by an astrophysical body, per unit time, in the range of electromagnetic waves.

MAIN SEQUENCE Set of stars that produce a continuous band in a plot showing the stellar light colour as a function of brightness (typically called Hertzsprung–Russell diagrams). These stars reach their stable state when the (outward) thermal pressure due to the heat produced by the internal nuclear reaction and the (inward) gravitational attraction equilibrate. The Sun is a typical star of a main sequence.

METRIC (see SPACE–TIME METRIC)

MICHELSON INTERFEROMETER Apparatus based on an optical device that measures the distance of luminous interference patterns, and can thus compare the length of an object to the wavelength of the employed light.

NEWTON Unit of force in the International System of Units: it is the force required to give a 1-kg mass an acceleration of $1 \, m/s^2$ (symbol: N).

OPEN (SPACE) Infinite nonflat space (as opposed to closed space).

PARSEC Unit of length equal to 3.26 light years (symbol: pc). Multiples are also used, such as the kiloparsec (kpc) for 1,000 parsecs and the megaparsec (Mpc) for 1 million parsecs.

PERIOD OF A WAVE Amount of time needed for a periodic phenomenon to repeat itself. The period is thus measured in seconds. Its inverse is the frequency.

PHASE TRANSITION Transformation of a system due to the change of a parameter of the system. For example, consider a system of matter and temperature as a parameter: phase transitions correspond to the transition from solid to liquid and from liquid to gas. In other systems, the energy (electroweak transition) or the amplitude of the magnetic field (ferromagnetic materials) can be considered as parameters.

PHOTODIODE Semiconductor component that converts light into an electric current.

PHOTON Elementary particle whose exchange between other particles is responsible for the electromagnetic force. The photon is also the fundamental constituent of light. It has zero mass and travels at the speed of light.

PLANET Celestial body orbiting around a star, with a sufficient mass for its own gravity to keep it in equilibrium in a quasi-spherical form.

PLASMA State of matter obtained at high temperature: the atomic or molecular bonds become broken and atoms (which are electrically neutral) become ions (electrically charged).

POLARIZATION Property of the waves which can oscillate with more than one orientation (such as electromagnetic or gravitational waves): each independent orientation is called a polarization of the wave. If the wave oscillates in only one of these orientations, it is said to be polarized.

POWERS OF 10 Notation which makes it possible to simplify the writing of huge or tiny numbers in the decimal system. For instance, 10^2 corresponds to 1 followed by 2 zeros, which is 100, 10^6 means 1 followed by 6 zeros, which is one million, and 10^{-2} corresponds to 0.01, which is 1 located at the second decimal, while 10^{-9}, located at the ninth decimal, means 0.000000001. And so on ...

QUADRUPOLE Distribution of mass, which is symmetrical with respect to rotations of $90°$ (it can be decomposed into four identical quadrants).

QUANTUM FIELD THEORY A theory that describes the quantum properties of fields. In particle physics this theory incorporates the formalism of *special relativity* and thus it describes particles/fields that are moving at speeds close (or equal) to the speed of light.

QUANTUM FIELD Field associated with a particle in the framework of quantum (and relativistic) theory. In particular, the field describes the fluctuations associated with the particle at each point in space and time: creation of a particle–antiparticle pair followed by its annihilation.

QUANTUM VACUUM Fundamental state (namely the state of minimal energy) for a region of space–time or for the whole Universe. The energy of this state is called vacuum energy. By introducing a particle with mass m in this region, the energy increases by a quantity mc^2.

RADIOMETER Device for measuring the intensity of the flux of electromagnetic radiation, in various frequency domains.

RANGE (OF A FORCE) Distance over which a force is exerted. This distance can be infinite (as is the case for gravitational and electromagnetic forces).

REFERENCE FRAME Frame of reference that specifies the exact position (with three numbers) and the exact time (with one number) of a given event. These numbers are called coordinates. The reference frame can be related to an observer (who is immobile in this frame), to Earth (which is immobile in this frame), or to a system of distant stars.

SCHWARZSCHILD'S RADIUS Radius associated (via the usual formula) with the area of the horizon of a Schwarzschild black hole, that is, the surface of the sphere in which matter and light are gravitationally trapped by a black hole.

SEMI-REFLECTIVE A body (or surface, as in the case of a mirror) that reflects only part of the incident light, while the rest of the light passes through it.

SPACE–TIME Four-dimensional continuum endowed with three spatial coordinates and one temporal coordinate, which enable us to localize any event or any particle at a particular instant.

SPACE–TIME METRIC Information given at each point of space and time about the local measure of distance and timespan: knowledge of the metric at a point of space–time (an event) yields the distance, or the time separation, to a nearby point.

SPECTRAL INDEX Number characterizing the dependence of flux of luminous energy on the frequency of its light.

SPECTRAL SHIFT Shift of the frequency (or wavelength) of the characteristic emission of a body due to its movement with respect to us (the visible light is red-shifted if the object recedes and blue-shifted if the object approaches us, according to the Doppler–Fizeau effect), or *alternatively* the number (denoted by z) that determines the ratio $(1 + z)$ between the wavelength of light detected on Earth and the emitted wavelength; this ratio also determines how much the Universe has expanded since the emission (e.g. the observable Universe today is $1 + 9 = 10$ times larger than at the time when a galaxy observed today with a spectral redshift $z = 9$ emitted its light).

SPIN Quantity that measures an intrinsic property of particles, similar to the mass or the electric charge. It enforces the rotational invariance of the laws of physics at a microscopic level. It is an integer number for bosons and a half-integer number (an odd number divided by two) for fermions.

STANDARD MODEL The theory describing the three fundamental interactions— electromagnetic, weak, and strong—at the level of elementary particles.

STAR An astronomical body that produces and emits energy.

TIDAL EFFECT Effect due to the *differential* gravitational attraction of a massive body by another one of finite size: the two bodies globally attract each other but, because of their finite size, the closest parts are slightly *more* attracted while the most distant are slightly *less*. The bodies can then be deformed, as is the case between the Moon and the liquid (thus deformable) mass of Earth's oceans. If the gravitational attraction is more significant (for instance, near a black hole horizon), these effects can lead to the rupture of a celestial body subjected to them.

TORUS Geometrical object that corresponds to a curved tube closed on itself.

TRANSPONDER Automatic device that receives, amplifies, and retransmits signals on different frequencies.

UNIFORM MOTION Constant-velocity motion.

VACUUM ENERGY Energy of the quantum vacuum due to quantum fluctuations in energy (production of virtual particle pairs).

VARIABLE STAR A star whose luminosity varies.

VIRTUAL PARTICLE Particle created (along with its antiparticle) from a quantum energy fluctuation. According to the laws of quantum mechanics, this fluctuation has a limited lifetime (the higher the energy, the shorter the lifetime), leading to annihilation of the particle with its antiparticle. The particle is virtual because it is not detected (otherwise, annihilation would not occur and the energy fluctuation would be permanent, in violation of the laws of energy conservation). Nevertheless, creation of virtual particles does have physical consequences (see VACUUM ENERGY, HAWKING RADIATION).

WAVE (see ELECTROMAGNETIC WAVE, GRAVITATIONAL WAVE)

Index

Page numbers in *italics* refer to figures.